Support Independent Type II

Support Independent Type II

Lars Harmsen
Marian Misiak
Pola Małaczewska

Slanted
2025

I

Introduction

We invite you on a remarkable journey into the fascinating, niche, and often overlooked world of type specimens—a unique crossroads of graphic design, typography, and storytelling. When we embarked on the creation of *Support Independent Type** in 2019–2020, our goal was modest: to bring together and celebrate the extraordinary creativity found in contemporary type specimens. What began as a heartfelt effort to showcase this specialized field grew into a much larger project than we could have imagined, sparking conversations, inspiring designers, and resonating across the global design community.

The first edition aimed to collect and present, in one place, a vibrant panorama of type specimens—a type of work rarely seen as a whole. It captured the imagination of designers, students, and typographic enthusiasts, with its striking content and bold manifesto for independent type. The overwhelming response to that book made it clear: this is only the beginning.

Since then, the landscape has continued to evolve. Over the past five years, we have witnessed an explosion of creativity in type specimens, with designers finding ever more inventive ways to promote and showcase new typefaces. These specimens are not just promotional tools; they are artifacts of innovation, reflecting the personality and artistry of their creators. With this second edition, we aim to shine a light on these developments and reaffirm our commitment to supporting the voices shaping the future of typography.

But what exactly do we mean by *independent type*?

At its core, *independent type* represents freedom—freedom to create, to innovate, and to express individuality. In typography, this translates to typefaces crafted by independent designers and small, passionate teams who operate outside the monolithic structures of corporate type foundries. These creators chart their own paths, offering type that is deeply personal, experimental, and sometimes provocative.

Independent type is a rebellion against uniformity. It is the antidote to the global homogenization of type design, where mega-corporations consolidate smaller foundries into sprawling, standardized libraries. In a world increasingly dominated by these giants, independent type stands as a bold counterpoint: diverse, dynamic, and unafraid to challenge the status quo. It thrives in the cracks between the mainstream, building an alternative that is as refreshing as it is necessary.

This second edition is not just a book; it is a declaration of support for those who prioritize creativity over conformity, innovation over imitation, and authenticity over assimilation. We believe that independent type is not only worth celebrating—it is essential for the vitality of the design world. Independent type designers are shaping the visual landscape of the future, pushing the boundaries of what typography can achieve.

With *Support Independent Type II*, we aim to inspire, educate, and advocate. We hope this expanded collection of type specimens—ranging from the intricately printed to the ingeniously digital—will serve as a testament to the power of independent creativity. By bringing these works together, we hope to amplify their voices and invite you to join this movement.

This project would not have been possible without the generous support of those who shared their time, resources, and expertise with us. To everyone who sent us printed materials, contributed digitally, or simply encouraged us with their enthusiasm—thank you. Your involvement is proof of the vibrant community spirit that makes independent type so extraordinary. With your help, we feel stronger, inspired, and even more determined to continue this journey. We are deeply grateful for this shared experience.

Independent type is more than just letters on a page; it's a statement, a community, and a vision for what design can be. Let's support it together.

**Support Independent Type. The New Culture of Type Specimens.*
L. Harmsen, M. Misiak, L. Panitz. Slanted 2020

II
Quotations

Can you imagine better conversion lubricant or public speech starter than with a bold and enlightening quote? In this chapter, we've gathered a collection of quotes about promoting typefaces and independent type foundries.
Typography may be visual, but the words used to describe and market it often carry as much weight as the fonts themselves. We explore how creators articulate their ideas, define their work, and sometimes craft unforgettable phrases to champion their designs. It's a showcase of witty, insightful, and occasionally poetic takes on type design.

Love via letters.

Fatype
fatype.com

Not all charac-
ters mean well.

About Odd Mono
Metis Foundry, metis-foundry.com

Drawing fonts is
my way of dancing,
singing, talking,
expressing myself.

Margot Lévêque Studio
margotleveque.com

undefined type undefined

NaN
nan.xyz

Letters carved in stones are our time travellers.

Gruppo Due
gruppo-due.com

You can just call it a font.

About Chateau
Astrae Studio, astrae.studio

Good typefaces are long-lasting design tools expressing style and feeling, helping designers to create great work.

Fatype
fatype.com

In type paradise, anything goes.

About Nirvana,
Altiplano, altiplano.xyz

Expert in nothing, with opinions on everything and successful in all kinds of failures.

FerCozzi
fercozzi.com

NN Didot Modern NN Didot Modern NN Didot Modern NN Didot Modern NN Didot Modern NN Didot Modern NN Didot Modern NN Didot Modern NN Didot Modern NN Didot Modern!

About NN Didot Modern
nouvellenoire.ch

Full industrial tech hacker vibes.

about Functor Mono
archetype-foundry.com

Kimera is Unity in Multiplicity.

Kimera
kimeracorp.eu

TEKIO is a home for creations focused around typography, sound, and vision.

Tekio
store.tekio.work

If you run around the rounded long enough, you might be inclined to make it square. And that's exactly what has transpired with the "Round" typeface.

About Round
bureaubrut.com

Coil feels comfortable like a well-worn pair of shoes.

About Coil
brownfox.org

Like technological innovations, we create typographic languages that express a world in motion.

Altiplano
altiplano.xyz

All Fonts Shall Behave.

DOUBLE ZERO
dblzr.com

From afar, all types might look the same. From up close, it's the details that do it.

Feedtype
feedtype.ca

J-LTF typefaces have other interests besides words.

J-LTF typefaces are biographical. They have had many lives and will have many more.

J-LTF typefaces care for their texts, amplifying the sentiments expressed by the content.

J-LTF typefaces differ in character, but are all steadfast in their solidarity.

Jung Lee Type Foundry
j-ltf.com

We believe in type as a vehicle for ideas, identity and expression.

Autograph
autograph.works

(...)We believe that many different narrations, shapes, beings can co-exist at different extremes of the design spectrum; and what is in-between might be an interesting surprise.

Blast Foundry
blast-foundry.com

Our process starts with a simple question: "Why is this typeface necessary?"

Lux Typo
luxtypo.com

The sans-serif typeface Denim fits any situation.

About Denim
Displaay Type Foundry, displaay.net

We could talk about:
The global impacts of monocultures.
Monocyclic developments.
But we'd rather talk about monospace.

About LL Riforma Mono
Lineto, lineto.com

NICE Poster—elegant, crisp, friendly. Best when served large.

About Nice
Jan Fromm, janfromm.de

We offer retail and custom typefaces with an approach based on research and conceptual thinking.

Kanon Foundry
kanonfoundry.com

Drawing influence and mood from phototypesetting era sans-serif fonts, Nicéphore is a disguised quiet guy.

About Nicéphore
Apex Type Foundry, apextypefoundry.com

Sometimes I'm thinking about quitting the internet but never TWK Everett.

About TWK Everett
Weltkern Typefaces, weltkern.com

Typefaces reflect a culture's priorities and aspirations. This is why we keep designing new typefaces—the job isn't done.

Klim Type Foundry
klim.co.nz

The SAANS, like Swiss typefaces, features a design that is universal and untitled, making it a default and common choice.

Displaay Type Foundry
displaay.net

We are dedicated to creating fonts that have something to say. Ones that capture voice, personality, and imagination, while altering perception of a brand or product.

Playtype
playtype.com

Bookish enough to be classy, peculiar enough to keep you coming back.

Source Type
sourcetype.com

Through intimately knowing the processes through which typefaces are made, we can produce work of better quality and functionality.

Mass-Driver
mass-driver.com

Capsule is the successful combination of humanism and technology: it embodies an artificial intelligence, smarter than artificial.

About Capsule
Altiplano, altiplano.xyz

Fun, Surprising, and Weird!

About Odd Mono
Metis Digital Type Foundry, metis-foundry.com

Hey we made a font and named it therma because it reminds us of thermal printer text and lo-fi machines.

About Therma
Astrae Studio, astrae.studio

After all, reminding of the past, the font talks about the modern times and it does it with those who have that very desire for renewal lit up inside.

About Tercia
Brownfox, brownfox.org

Akeed is a conventional arabic display font in a contemporary context, floats like a panda, stings like a dinosaur.

About Akeed
Hey Porter, heyporterposter.com

Our design is solid and clear, sometimes minimal, others bolder and more irreverent. We prefer irony to taking ourselves too seriously.

Due Studio
due-studio.com

We believe that art practice integrated with the exploration of letterforms (and a little humour) can provoke a refreshing wave of display typography.

About Rubin Sans
Threedotstype, threedotstype.com

GT Ultra dances between the worlds of Sans and Serifs, fusing calligraphy and construction.

About GT Ultra
Grilli Type Foundry , grillitype.com

In Gradient, we strongly believe that the deconstruction of traditional approach in typography combined with various type experiments, can re-imagine the future of contemporary typography.

Gradient
wearegradient.net

We really feel that our fonts should be shared, used, and viewed by as much people as possible.

Pangram Pangram
pangrampangram.com

Shriftovik isn't just a type foundry, but it's a team of enthusiasts who believe that fonts change the world.

Shriftovik Type Foundry
shriftovik.com

We believe that convincing communication is only possible with the most aesthetically pleasing and most technically sound fonts.

Fontwerk
fontwerk.com

Working with vernacular references and from intrinsic ideas, our typefaces are designed to have their own reason for being, rather than filling a gap in the font market.

Or Type
ortype.is

Light as a feather, elegant and sharp.

About Loretta Display
Nova Type Foundry
novatypefoundry.com

Our practice stems from research, friendship, thoughts, and memories experienced through fieldwork and active investigation.

Supercontinente
supercontinente.com

Small but mighty—as the adage goes!

About Tiny Gothic
Optimo Type Foundry, optimo.ch

Nuform is a type design practice that looks to the culturally offbeat and peculiar oddities to inspire and inform the creation of expressive letterforms.

Nuform Type
nuformtype.com

Our typefaces are drawn with care and explore themes of authorship, archetypal and emerging letter models, and functionality.

Road to Venice Type
r-vtype.com

To us, Heheh is a playground, a space for collaborative experimentation and support.

Heheh Type
hehehtype.com

British Standard Type develops progressive and distinctive typefaces that move the type design discourse forward.

British Standard Type
britishstandardtype.xyz

—II—

Printed Specimens

Printed type specimens are the heart of this publication. Over the years, this field has seen remarkable progress in both creativity and innovation. These specimens are more than promotional materials—they are a masterclass in editorial design and a treasure trove of information about typefaces and their unique features. Think of them as the haute couture of typography: meticulously crafted, stunningly beautiful, and timeless in their appeal. They're hard to resist collecting.

Va
gl
ar
st
part
21 03 21 17 04 UTC+1
CH-TWK
NP
IM
JC

on
of
today
of
particular
/
culture
including
specificity,
to
I'm

Everett Hairline Italic
Everett Thin Italic
Everett Ultralight Italic
Everett Light Italic
Everett Regular Italic
Everett Medium Italic
Everett Bold Italic
Everett Extrabold Italic
Everett Black Italic
Everett Super Italic
Everett emerged during Nolan Paparelli's studies at ECAL/University of Art and Design Lausanne, culminated in his diploma project in 2015 and has been steadily evolving ever since. Initially inspired by the work of American photographer Daniel Everett, in particular his architectural collages, the project soon became more personal and reflects the designer's own vision of the grotesque genre. Everett's identity is marked with distinctive cuts on apexes and vertexes of letters containing diagonal strokes. These powerful typographic effects add a great tension to the letters individually, while at the same time ensuring a good reading comfort in running texts, finding the right balance between a font that is graphic yet fluid. The typeface's pronounced cuts are compensated with an organic drawing, smooth curves, open junctions in the stems, relatively high stroke contrast and regular proportions. Set in large sizes, the typeface shows a strong graphic impact while when reduced to smaller sizes, these effects fade away and the typeface approaches the aesthetics of a traditional workhorse neo-grotesque. The result of this design proves a real versatility which is rare in this particular genre of sans serif. The vertical dimensions of the typeface allow also users to compose texts with tight line spacing, thus making it possible to save space on the page. The variation of weights from Hairline to Super with corresponding italics form a large and coherent font family consisting of 20 styles, offering users many possibilities in various design applications, from books to posters, from branding to signage, on- and offline. Each weight has been designed differently in relationship to its own tension, resulting in the characteristic cuts being more open and airy in the lighter weights while the strokes meet again and become more compact in bolder weights. Last but not least, Everett also includes many OpenType features (stylistic sets, ligatures, superiors/inferiors, fractions, etc.), giving designers the freedom to work with a complete and versatile tool in the scope of their graphic design projects. Constantly oscillating between display and text, expression and tradition, mechanical and organic, Everett is resolutely a contemporary neo-grotesque typeface with distinctive features that demonstrate its daring, striking character.
TYPEFACE
Nolan Paparelli
Igino Marini
Jan Charvát
TYPE.WELTKERN®
21 March 2021
SPECIMEN
Nolan Paparelli
Daniel Everett
Offizin Scheufele
WELTKERN®
March 2022

Lift Type
Typeface: Baste A & Baste B
Design: Image Format
—
lift-type.fr @lifttype

Maxitype
Typeface: OCR-X
Design: Eurostandard, Maximage
—
maxitype.com @maxitype.com

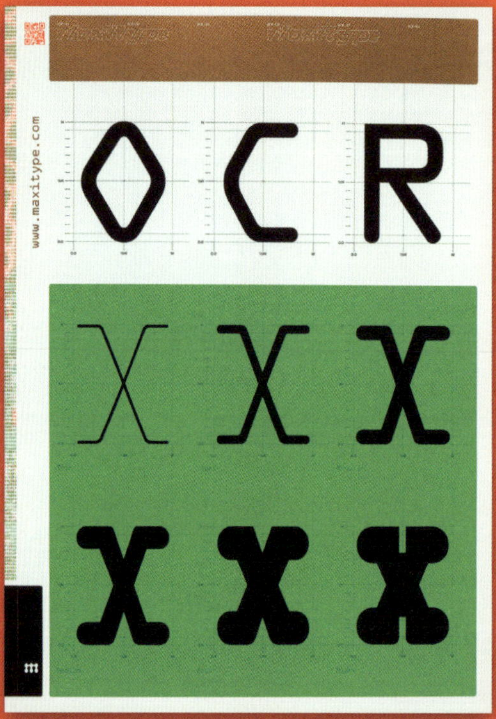

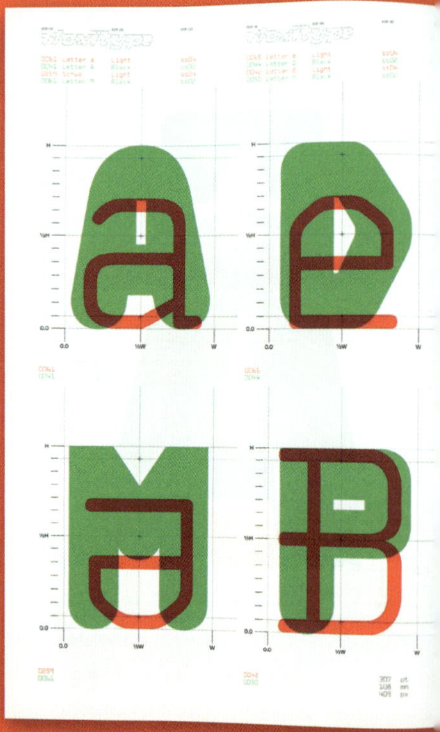

OCR-X

Jonas Berthod 1/3

ONLY COOL RIDE

Designed by Eurostandard and Maximage, OCR-X is both a homage to and an extension of the infamous OCR-A typeface from 1966. Taking its predecessor's skeleton as a starting point, OCR-X reinterprets it and expands it in six weights. The Regular is a revival of the original cut of OCR-A offering lifted details and variations. For instance, aligned punctuation and alternative glyphs are available as stylistic sets. The Thin cut underlines the hexagonal and geometric construction of the OCR-A system, while the Bold and Black morph into a new, blocky variation. The versatility across the cuts means that OCR-X can be used on both screen and print. Its extended character set makes it a highly functional typeface, fully aligned with today's standards. A variable version of the typeface is also available on request.

ALIEN STANDARDS

OCR-A was created in 1966 (1) by the American Type Founders (ATF) (2) on a commission by the American National Standards Institute (ANSI, the North American equivalent of the Deutsches Institut for Normung – DIN – (3) to whom we owe standardized paper formats). The ANSI required a highly differentiated monospace to ensure successful recognition with the comparatively low-resolution optic readers available in the 1960s. This restrictive logic determined OCR-A's unusual letter shapes, evidenced for instance in its lozenge-shaped O, skewed Q, kooky C, flat S and unusually high M. Its construction was meant to be recognizable on a small optical grid of five by nine units. Its orthogonal connections and rounded ends, but also its peculiar punctuation (including blocky symbols) all make it seem like it was designed either by a cyborg intern from Star Wars' Yavin 4 or by someone taking part in the Space Race (who I like to imagine was wearing one of Paco Rabanne's metal dresses).

OCR-A's numerals also draw from its predecessor E13B (1958), which was developed over ten years at the Stanford Research Institute under the direction of Kenneth R. Eldredge after a commission by the American Bankers Association to develop an automated cheque-processing system. E13B's highly unusual shapes were determined by the functionality of magnetic ink readers. Each glyph is drawn to be read on a seven-by-ten square matrix and includes peculiar protuberances that allow characters to be recognized even when folded, misaligned, or partially missing. The typeface is designed to be printed with magnetic ink at the bottom of cheques. Reflecting its use, the character set only includes the ten digits and four processing symbols (transit, amount, on-us and dash), in contrast to OCR-A, which contains a full character set of ninety-six core glyphs (plus nine national letters and currency signs), all available in three sizes (I, III and IV respectively) to accommodate different brands of readers.

1 – Though some sources date OCR-A to 1968, the foreword to the ANSI X3.17-1981 standard confirms that it was "first developed in the United States in 1961 as a numeric font only. In 1966 an alphanumeric font which contained 57 characters, including the existing numeric font, 4 abstract characters, and only capital letters, was issued. The revised standard was entitled American National Standard Character Set for Optical Character Recognition, ANSI X3.17-1966."

2 – The ATF company (not to be confused with the ATF association) was founded in 1892 to consolidate twenty-three of the most prominent type foundries in the US, which felt threatened by the recent arrival of automated mechanical typesetting machines, the Linotype and the Monotype.

3 – Like the ANSI, the DIN enshrined OCR-A in a norm: DIN 66008.

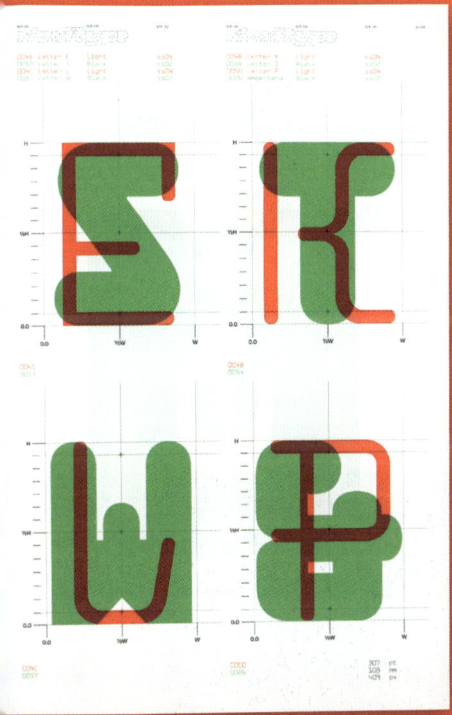

Rebekka Hausmann
Typeface: Rank
Design: Rebekka Hausmann
—
rebekkahausmann.de @rebekkahausmann

Support: Independent, Type II

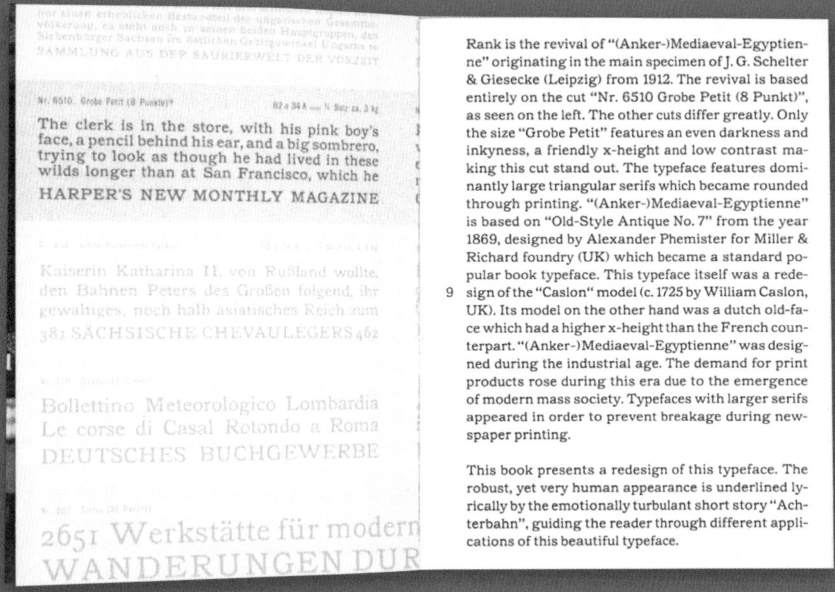

Nr. 6510 Grobe Petit (8 Punkte)*

The clerk is in the store, with his pink boy's face, a pencil behind his ear, and a big sombrero, trying to look as though he had lived in these wilds longer than at San Francisco, which he
HARPER'S NEW MONTHLY MAGAZINE

Kaiserin Katharina II. von Rußland wollte, den Bahnen Peters des Großen folgend, ihr gewaltiges, noch halb asiatisches Reich zum
381 SÄCHSISCHE CHEVAULEGERS 462

Bollettino Meteorologico Lombardia
Le corse di Casal Rotondo a Roma
DEUTSCHES BUCHGEWERBE

2651 Werkstätte für modern
WANDERUNGEN DUR

Rank is the revival of "(Anker-)Mediaeval-Egyptienne" originating in the main specimen of J. G. Schelter & Giesecke (Leipzig) from 1912. The revival is based entirely on the cut "Nr. 6510 Grobe Petit (8 Punkt)", as seen on the left. The other cuts differ greatly. Only the size "Grobe Petit" features an even darkness and inkyness, a friendly x-height and low contrast making this cut stand out. The typeface features dominantly large triangular serifs which became rounded through printing. "(Anker-)Mediaeval-Egyptienne" is based on "Old-Style Antique No. 7" from the year 1869, designed by Alexander Phemister for Miller & Richard foundry (UK) which became a standard popular book typeface. This typeface itself was a redesign of the "Caslon" model (c. 1725 by William Caslon, UK). Its model on the other hand was a dutch old-face which had a higher x-height than the French counterpart. "(Anker-)Mediaeval-Egyptienne" was designed during the industrial age. The demand for print products rose during this era due to the emergence of modern mass society. Typefaces with larger serifs appeared in order to prevent breakage during newspaper printing.

This book presents a redesign of this typeface. The robust, yet very human appearance is underlined lyrically by the emotionally turbulant short story "Achterbahn", guiding the reader through different applications of this beautiful typeface.

9

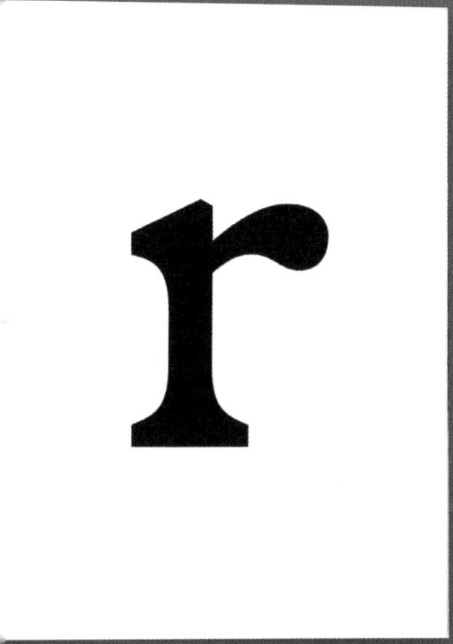

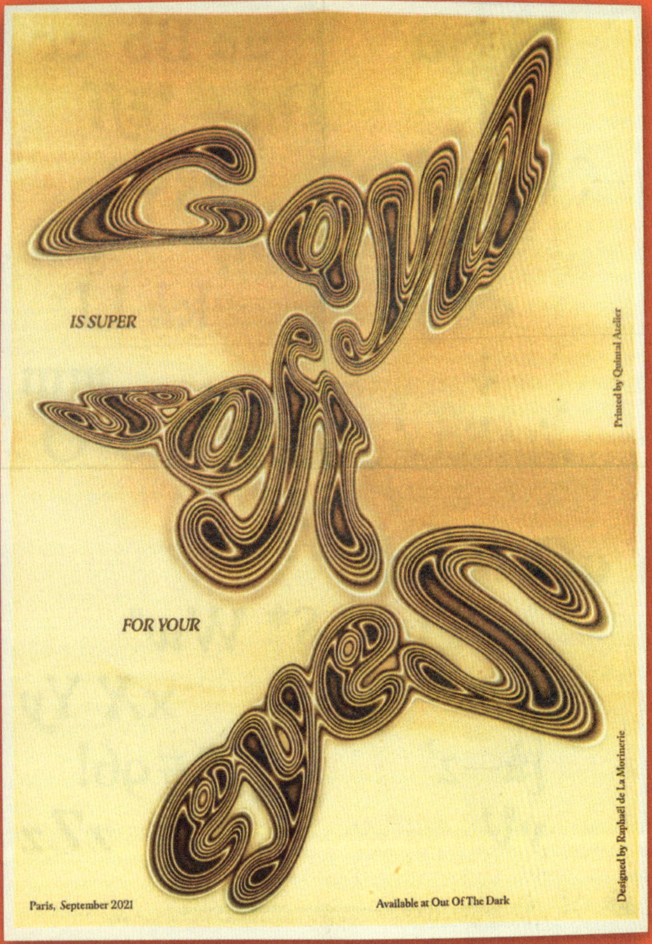

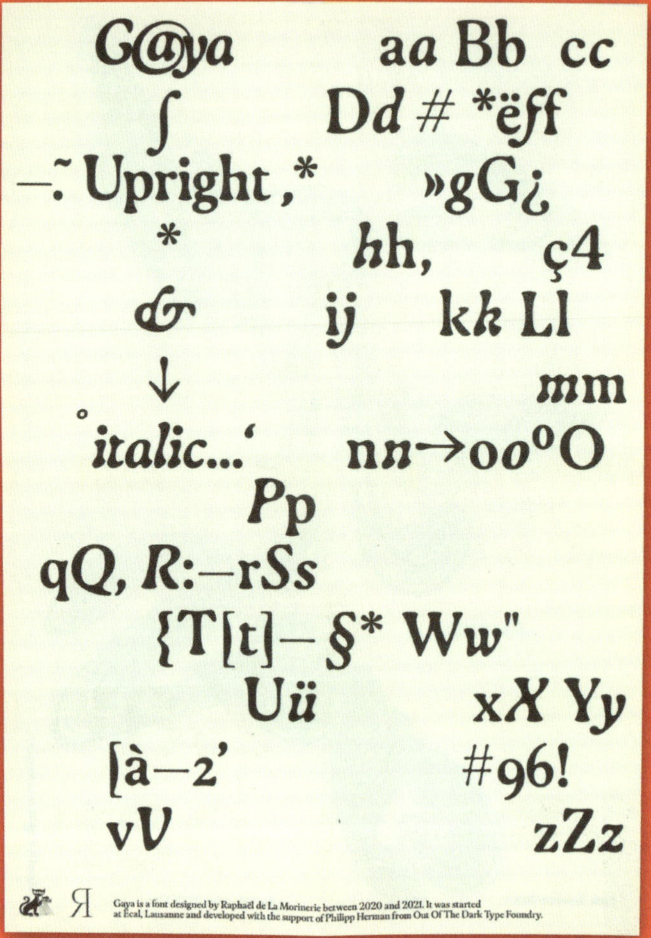

Out of the Dark
Typeface: Gaya
Design: Raphaël de La Morinerie
—
outofthedark.swiss @outofthedark.xyz

Lift Type
Typeface: Sans Plomb
Design: Image Format
—
lift-type.fr @lifttype

SansPlomb®
Sans
Sans
Sans
Formule®
400
IMAGEFORMAT.FR
LIFT-TYPE.FR
Speedlong®
IFRA

Support Independent Type II

Lift Type
Typeface: Sans Plomb
Design: Image Format
—
lift-type.fr @lifttype

Lift Type
Typeface: Sans Plomb
Design: Image Format
—
lift-type.fr @lifttype

Support Independent Type II

Jeremy Tankard Studiotype
Typeface: Several
Design: Jeremy Tankard
—
studiotype.com @jeremytankardtypography

Maxitype
Typeface: Selecta
Design: Maximage (Daniel Haettenschwiller)
—
maxitype.com @maxitype_com

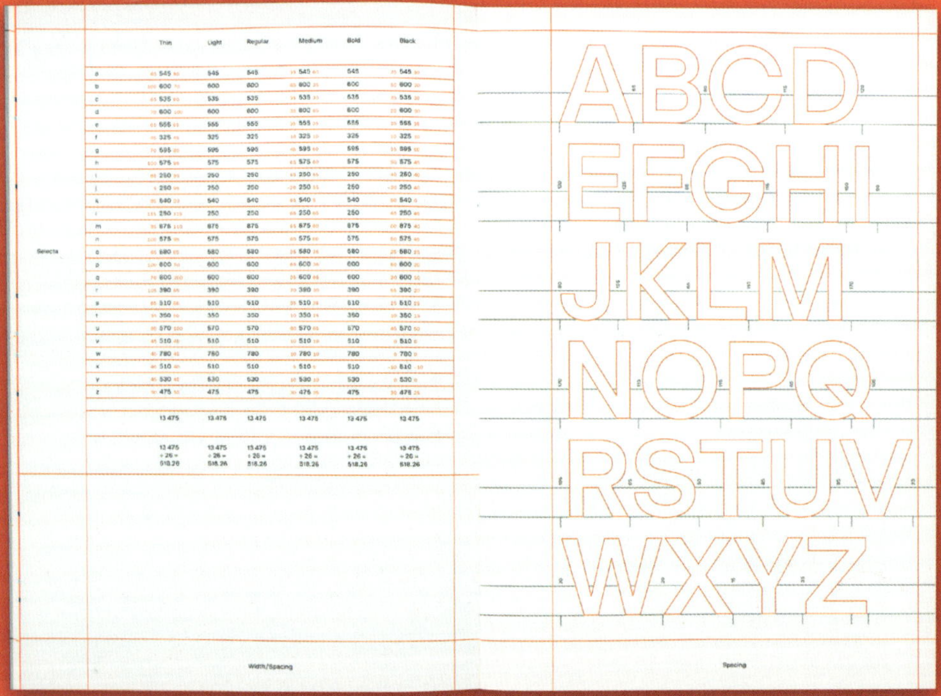

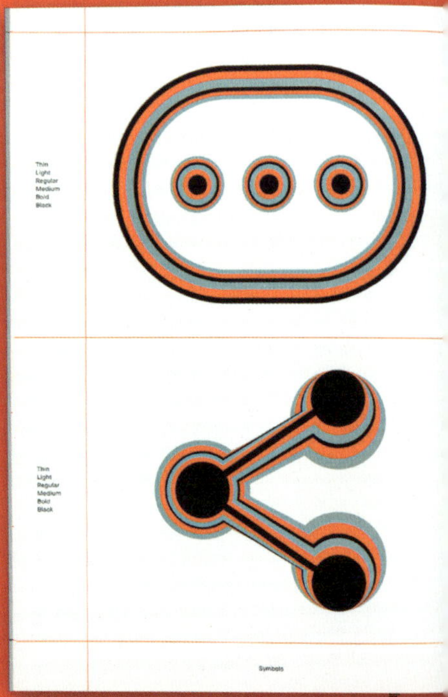

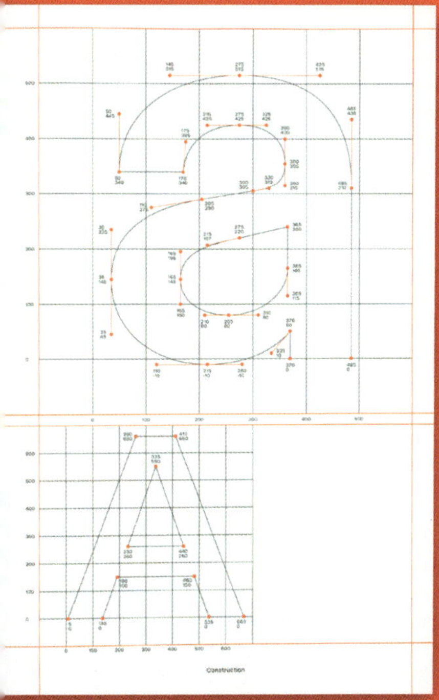
Construction

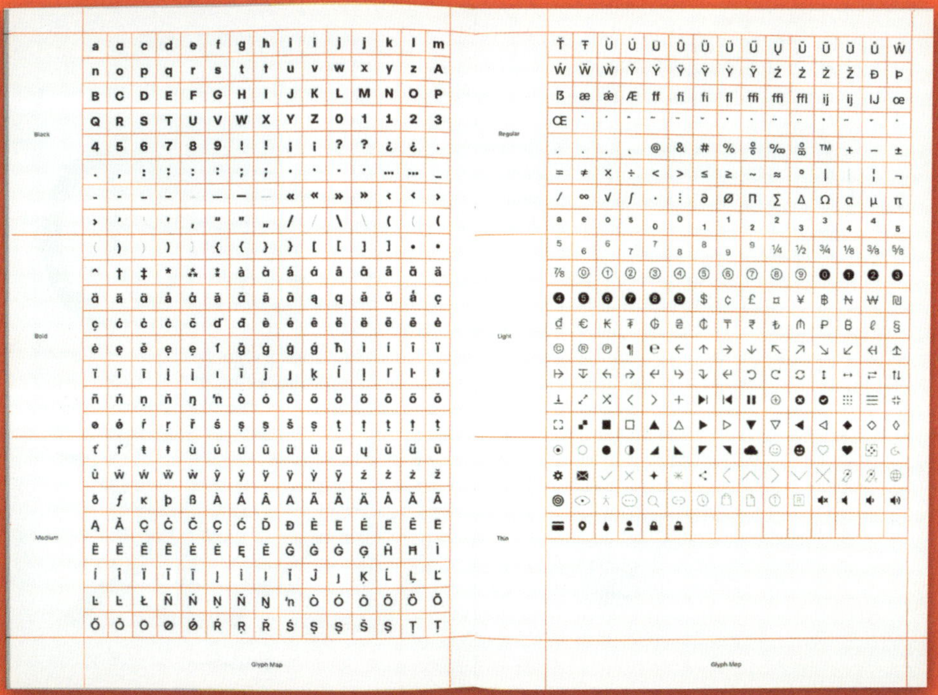
Glyph Map
Glyph Map

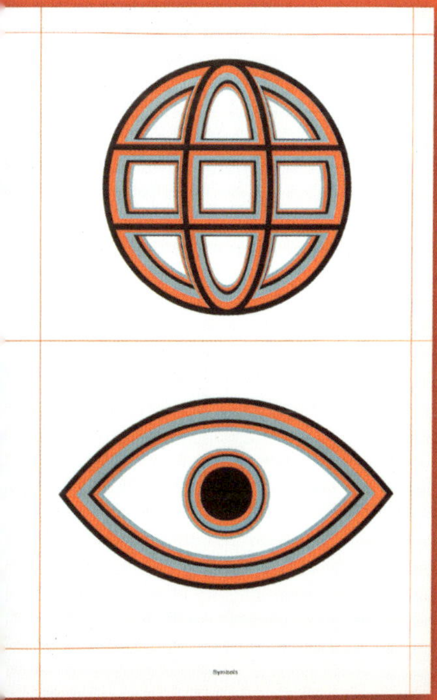
Symbols

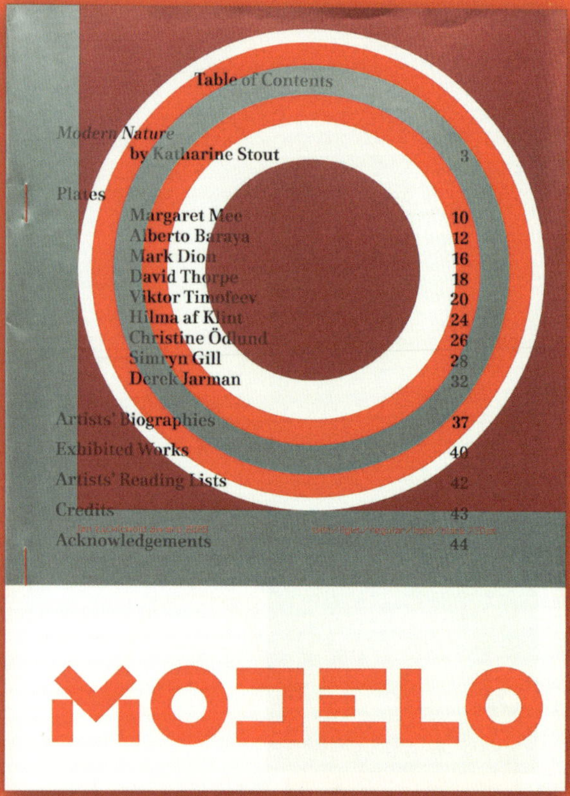

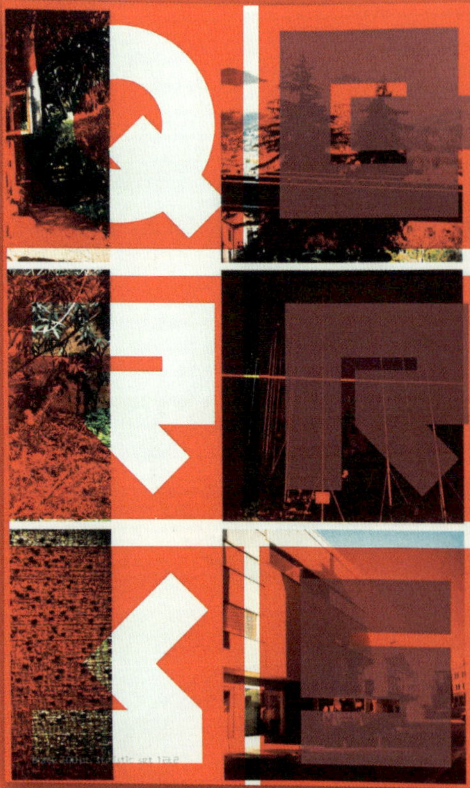

Maxitype
Typeface: Modelo
Design: Maximage (Daniel Haettenschwiller)
—
maxitype.com @maxitype_com

Support Independent Type II

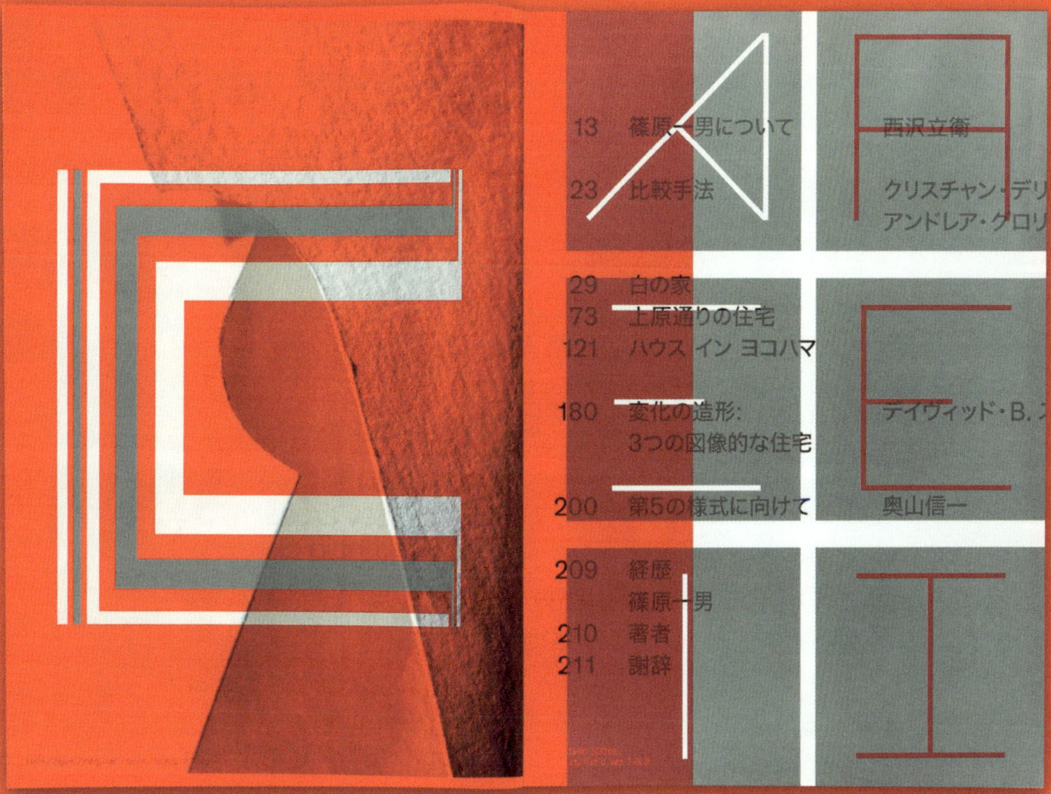
13 篠原一男について 西沢立衛
23 比較手法 クリスチャン・デリ
アンドレア・グロリ
29 白の家
73 上原通りの住宅
121 ハウス イン ヨコハマ
180 変化の造形:
3つの図像的な住宅 デイヴィッド・B.
200 第5の様式に向けて 奥山信一
209 経歴
篠原一男
210 著者
211 謝辞

A B C D
X Y Z !?
bold 120pt

FREE
FONT
black 120pt
www.maxitype.com

bb-bureau
Typeface: Pimpit
Design: Benoît Bodhuin
—
bb-bureau.fr @benoitbodhuin

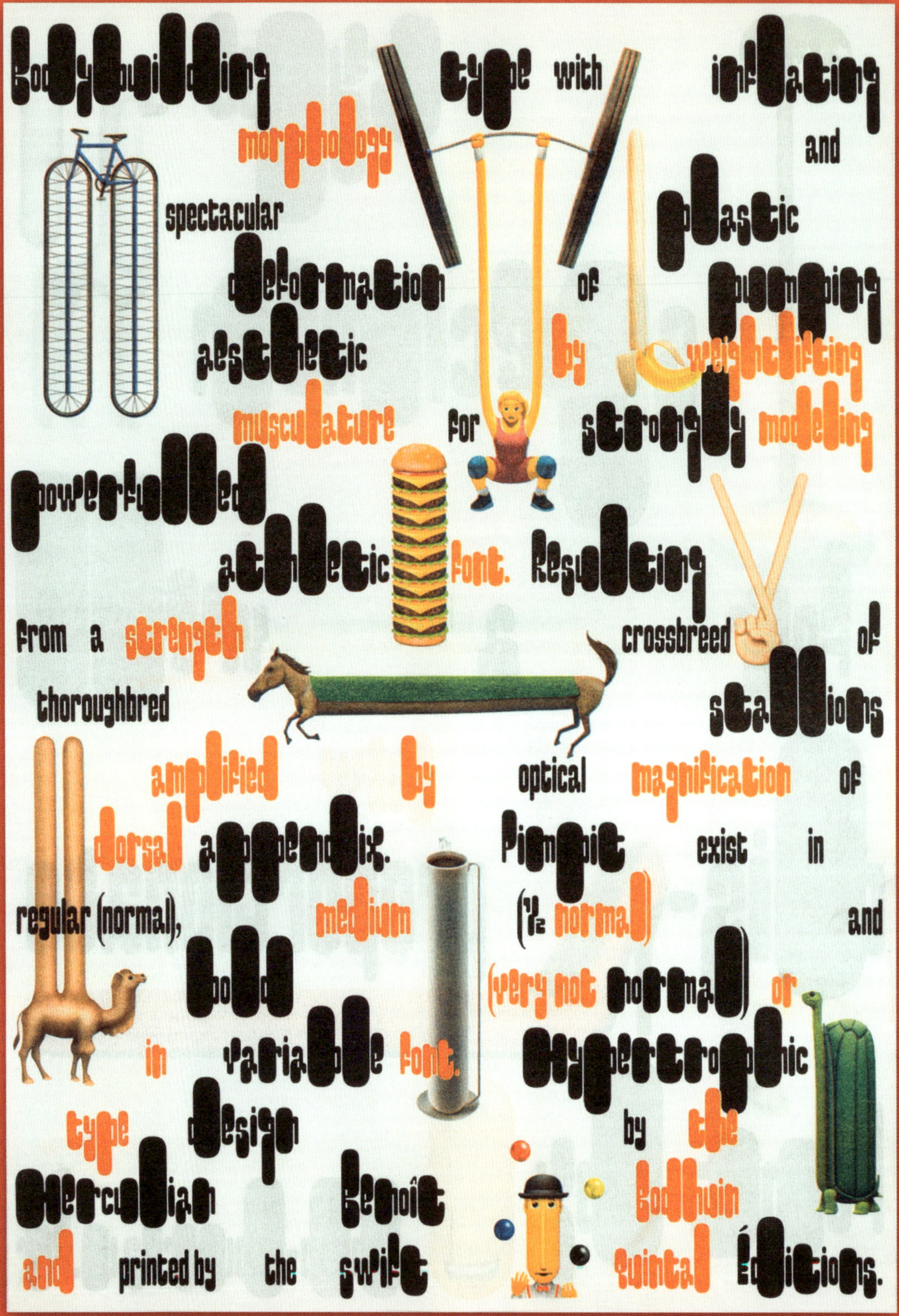

Support Independent Type II

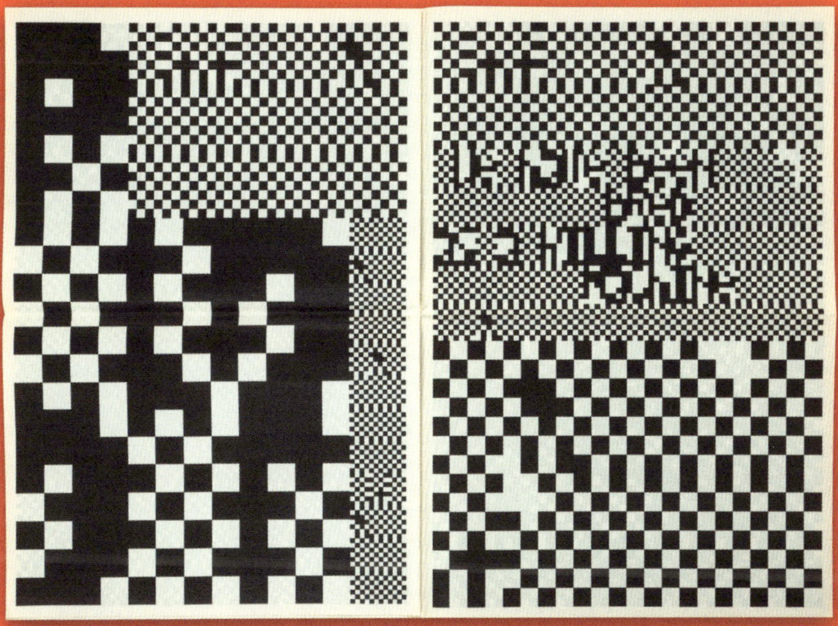

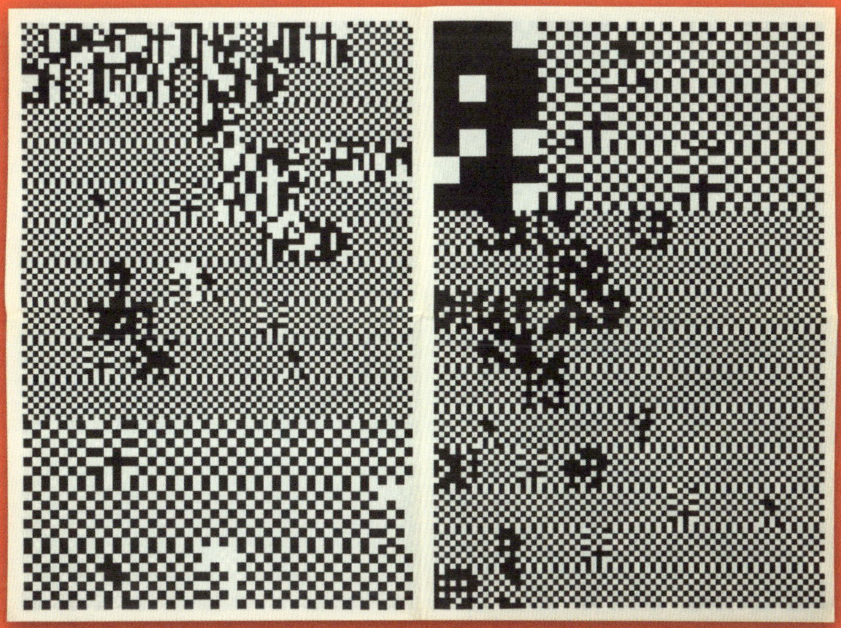

Simon Renaud
Typeface: Cipher E
Design: Simon Renaud
—
simonrenaud.fr @smnrnd

bb-bureau
Typefaces: Tokyto, Harber, Gikit, Moki
Design: Benoît Bodhuin
—
bb-bureau.fr @benoitbodhuin

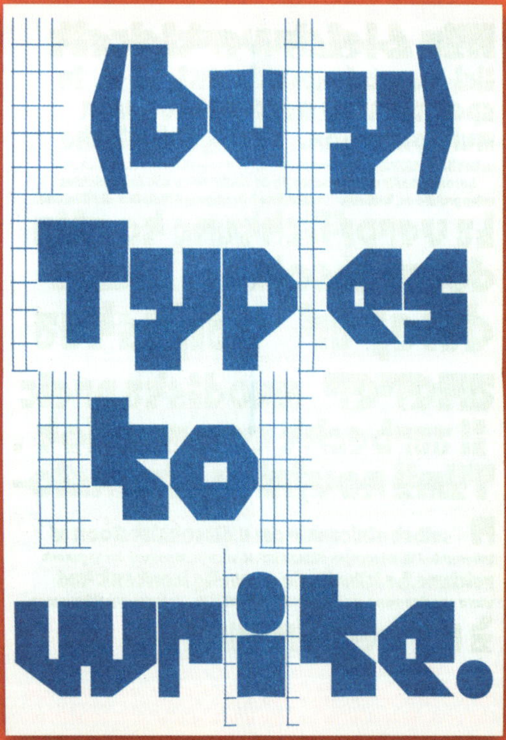

Support: Independent. Type II

HARBER is a rounded type designed of dots on a grid. Letters are invariable, only dots change, parameterized by five axis: weight, slant, volume, noise and optical size. Type and specimen designed by Benoît Bodhuin. Specimen printed by Quintal Éditions.

Type design
Parameterized
optical size
Weight
Editions

Moki & Moki
(type)
& (specimen)
are designed by
Benoit Bodhuin
and printed by
Quintal-atelier

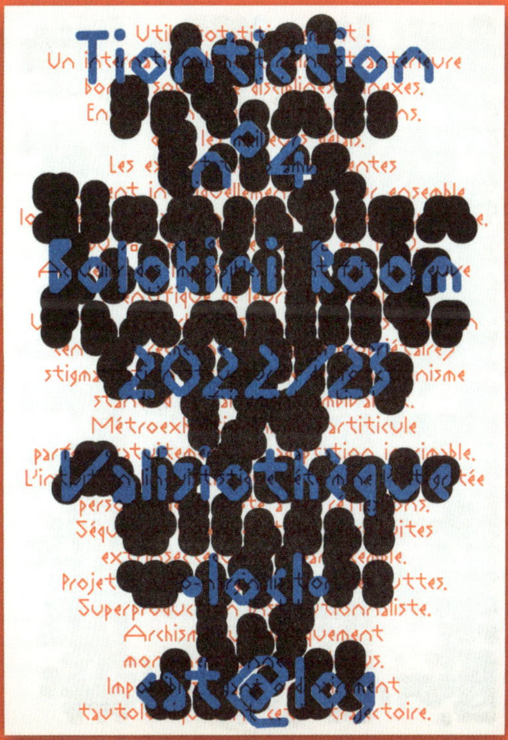
Tiontiction
n°4
Solokini Room
2022/23
Valisiothèque

Dinamo Typefaces
Typeface: ABC Arizona
Design: Elias Hanzer
—
abcdinamo.com @abcdinamo

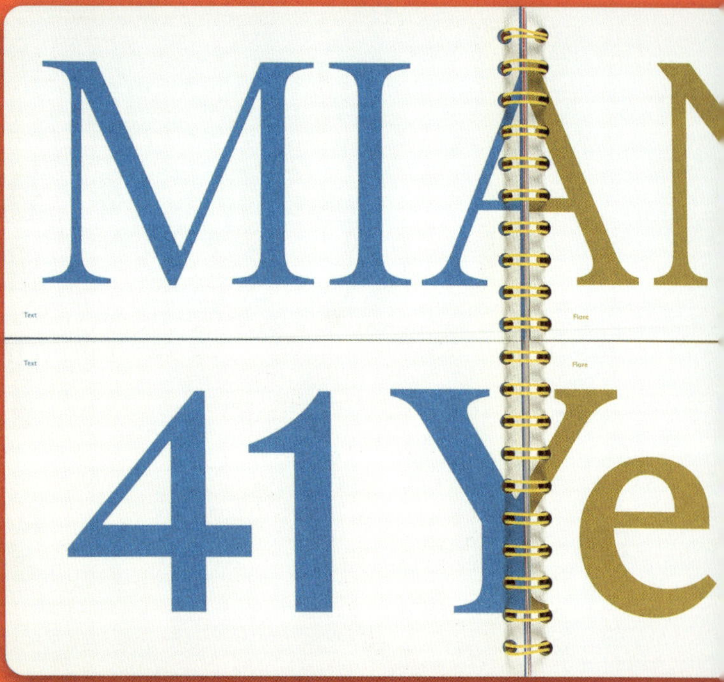

NS
ar
E
OPEN TYPE FEATURES
1 [ss01] Alternate single-story a
2 [ss02] Alternate single-story g
3 [ss03] Alternate punctuation
4 [ss04] Alternate Italics
5 [ss05] Alternate 1 and 4
6 [ss06] Alternate Q, g and &
7 [ss07] Alternate a
ALT Alternates
TF Tabular Figures
OSF Tabular Figures
CS Case Sensitive
SL Standard Ligatures
Æ
Light, Bold 14pt
herself had joined the team from the H. Berthold type foundry, as did Klein, who was already familiar with Ikarus. The concept of joining sans and serif in a shared design space had already been explored by Jan van Krimpen in the late 1930s, and Rotis is also predated by several contemporaries, including 1985's Lucida by Charles Bigelow and Kris Holmes, which includes serif, sans, typewriter, blackletters, and several other styles.
THE ADAPTATION OF INTERPOLATION TO TYPE DESIGN WAS A CRUCIAL STEPPING STONE IN THE DEVELOPMENT OF VARIABLE FONT TECHNOLOGY (VF) IN 2016. VFS KEEPS CONCEPTS OF LARGE TYPE FAMILIES VERSATILE IN RANGE AND APPLICATION, AND COMPRESSES MULTIPLE WEIGHTS OF DIFFERENT STYLES IN A SINGLE FONT FILE. ARIZONA ENTERS THE EXQUISITE CIRCLE OF MULTI-GENRE TYPE FAMILIES: ITS SPECTRUM EXTENDS FROM SANS, A HUMANIST-FLAVORED ANGLO-AMERICAN GROTESQUE, TO SERIF, A TRUE RENAISSANCE MODEL THAT RETAINS CONTEMPORARY FRESHNESS THE THINNER IT GETS. ARIZONA ALSO BALANCES THE DISTINCTLY-SERIFED COPPERPLATIAN FLARE AS WELL AS THE MIX, A FACE THAT KEEPS ITS EPONYMOUS PROMISE WITHOUT FALLING INTO SOME KIND OF POST-MODERN AWKWARDNESS. EACH MEMBER EASILY OFFERS A STAND-ALONE SOLUTION, BUT THE TRUE ESSENCE LIES IN THEIR CONVERGENCE, WITH EVERY INSTANCE BETWEEN THEM AT HAND, SHAPED BY TOOLS AND TECHNOLOGY.
HOBGOBLIN, Goblin
ALP, Pathfinder
BASILISK, Cross-Breed
Q&
Serpent-like dragon that twined and twisted around the tree in the Garden Hesperides and guarded the Golden Oranges.
Serif
Medium 145pt
Regular, Italic 20pt
THE
Griffin
Italic, Regular 150pt
Serif
Bold, Bold Italic, Regular, Italic 20pt
It stops in front of a house and bellows three times.
Conceived of as the ghosts of the arrogant!
About how one would hear a laughter out of nowhere?
By devils who lay two lit candles on its horns,
and prod it with hot skewers to exorcise the spirits...
Light, Medium 15pt
Was the name of a devil or daemon haunting the country around Évreux, Normandy. Alternatively, it may be a diminutive or other derivative of the old French proper name - type of knocker, derives from extinct languages.
Bold, Light 60pt
Bardha
Qiqirn
Thin, Thin Italic 95pt
[GARUDA]
(BIRD)
[PEOPLE]

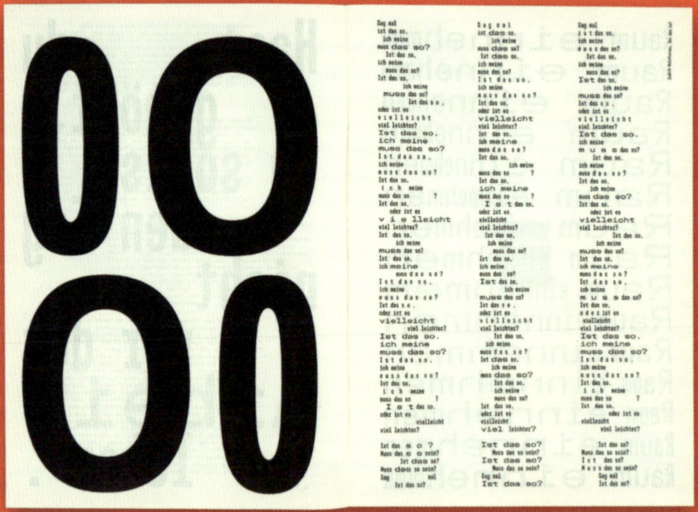

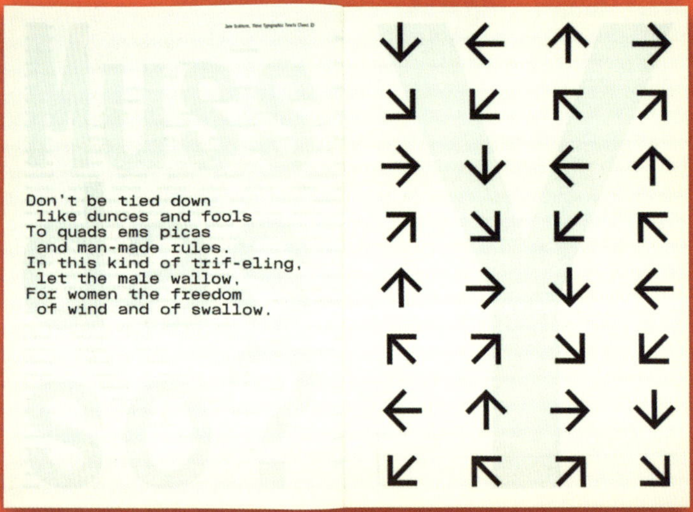

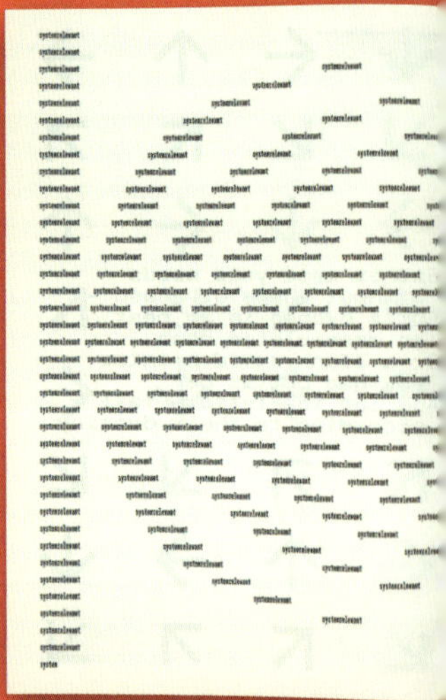

Lea Johanna Becker
Typeface: Nomo Mono
Design: Lea Johanna Becker
—
leajohannabecker.com @leajohannabecker

ist dieses morgen denn
ein leben ohne h e u t e wert?
ist es das wert.
dass man sich nie
gegen
die meute
w e h r t ?
sowieso
so
wie
so?
wieso
so ?
Ist
das
so?
nomomono

GOOD
LUCK.

Pizza Typefaces
Typeface: Kern
Design: PizzaTypefaces
—
typefaces.pizza @pizzatypefaces

WELCOME TO
FEAR CITY
A Survival Guide for Visitors to the City of New York

®HA
LUC
#CO
SA
IN P
2PM
№
911
23L

bold capital words: WELCOME TO FEAR
different era, and the pamphlet was pa
away–all in retaliation for a slew of layo
other public agencies.
ANDBAG
GGUAGE
CKTAILS
FEGUARD
UBLIC⇆
M FIRMLY
SECURE
DEPT.NY
LOUNGES
WELCOME TO
FEAR CITY
Kern Typeface is available
only on ty

46 pt

. This was the New York City of a
a Council scare tactic to keep tourists
hat shrunk the city's police force and

Good luck!

NEW YORK CITY 1975

atic & Variable font format

faces.pizza

Nouvelle Noire
Typeface: Swinton
Design: Marc Droz
—
nouvellenoire.ch @nouvellenoire_ch

HAW Hamburg
Typeface: Octagon Variable
Design: HAW Hamburg
—
haw-type-design.github.io

Support. Independent. Type II

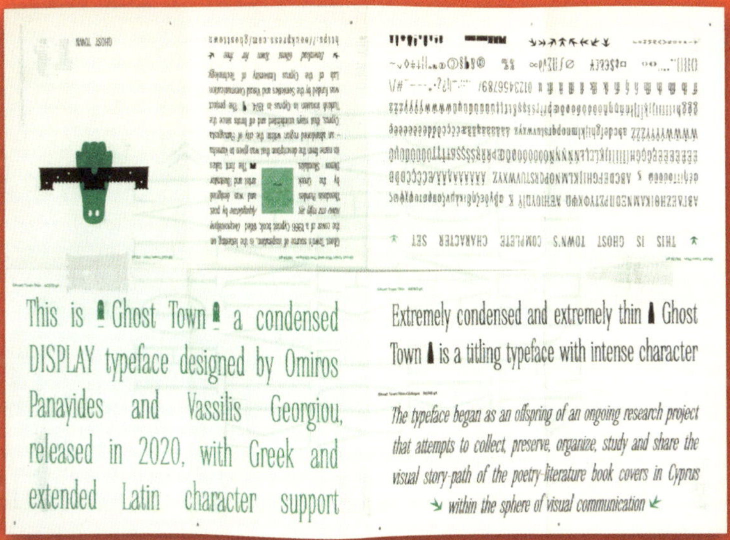

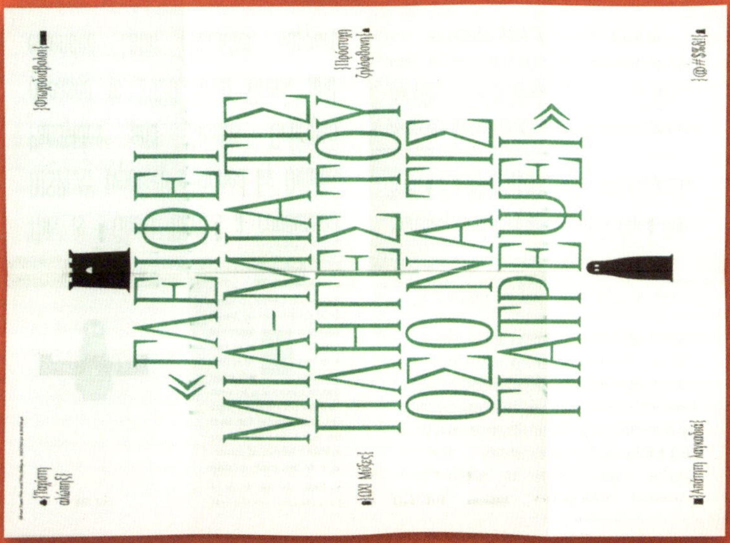

Omiros Panayides
Typeface: Ghost Town
Design: Omiros Panayides, Vassilis Georgiou
—
omirospanayides.com @omrspnyds

A TO Z WITH BISEL
Gloves
Quartz
Ravioli
Chair
Drone
BISEL IS A VARIABLE TYPEFACE DESIGNED BY FASE STUDIO FOR THE DESIGNERS FOUNDRY.COM
@FASE.STUDIO @TDFOUNDRY
ISBN 978-0-473-69262-9

Hwawon
Shin-Yang-Jang-Phonogram
Design: Shin Donghyeok (Specimen), Donghyeok Shin (Design)
—
@hwawon_mediabus

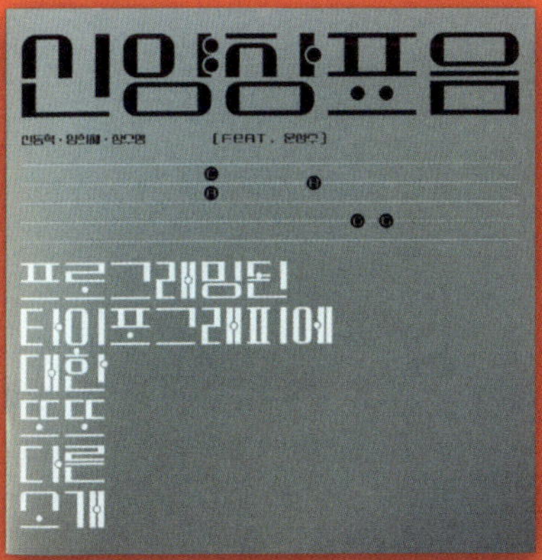

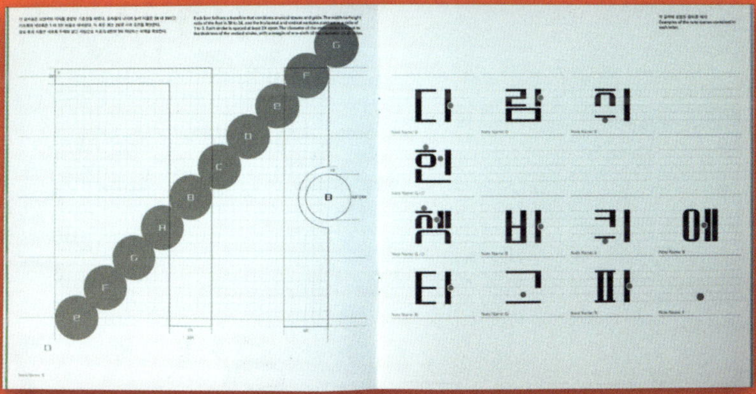

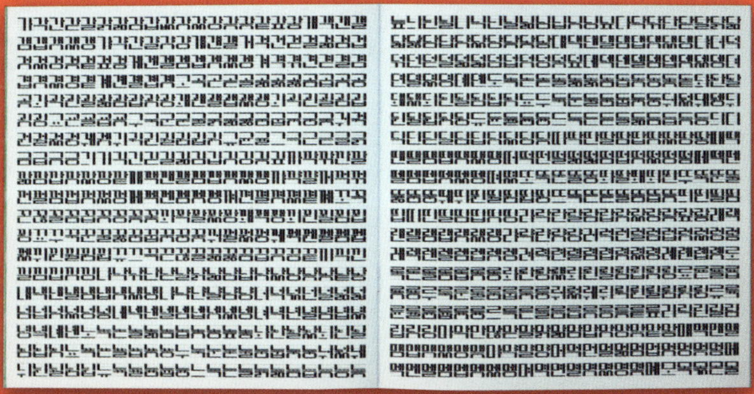

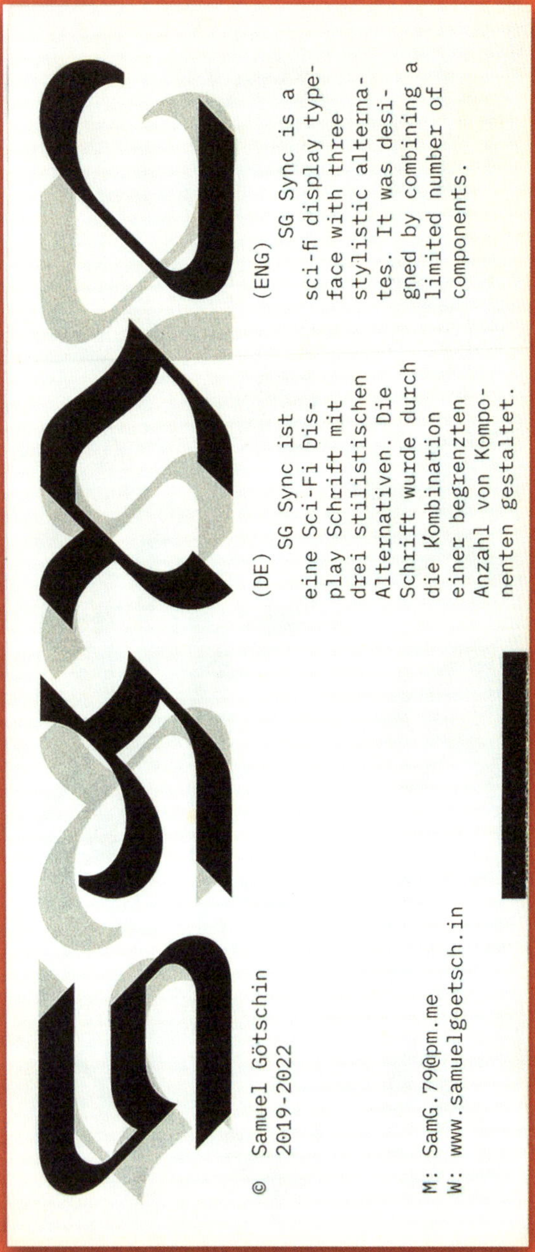

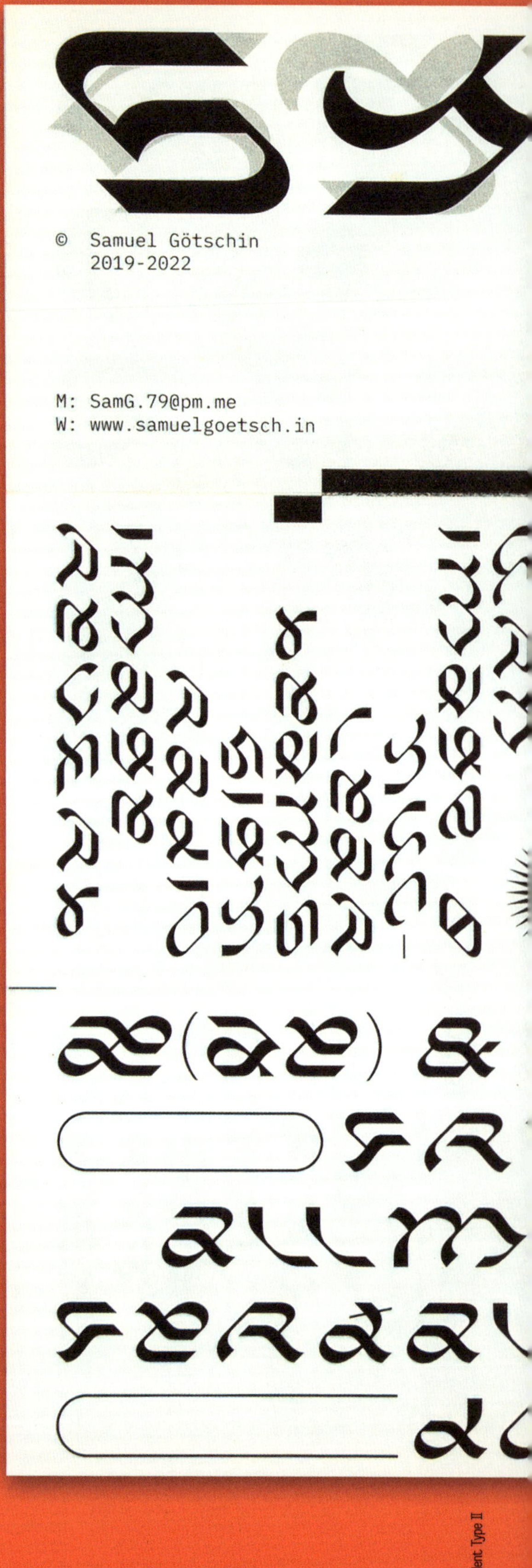

Samuel Götschin
Typeface: SG Sync
Design: Samuel Götschin
—
samuelgoetsch.in

(DE)

SG Sync ist eine Sci-Fi Display Schrift mit drei stilistischen Alternativen. Die Schrift wurde durch die Kombination einer begrenzten Anzahl von Komponenten gestaltet.

(ENG)

SG Sync is a sci-fi display typeface with three stylistic alternates. It was designed by combining a limited number of components.

(DE)

Funktionen & Eigenschaften:

- Kontextbedingte Glyphenvarianten
- Europäische Sprachunterstützung
- Unicase (A=a)

(ENG)

Features & prope

- Contextual stylistic
- European langugae su
- Unicase (A=a)

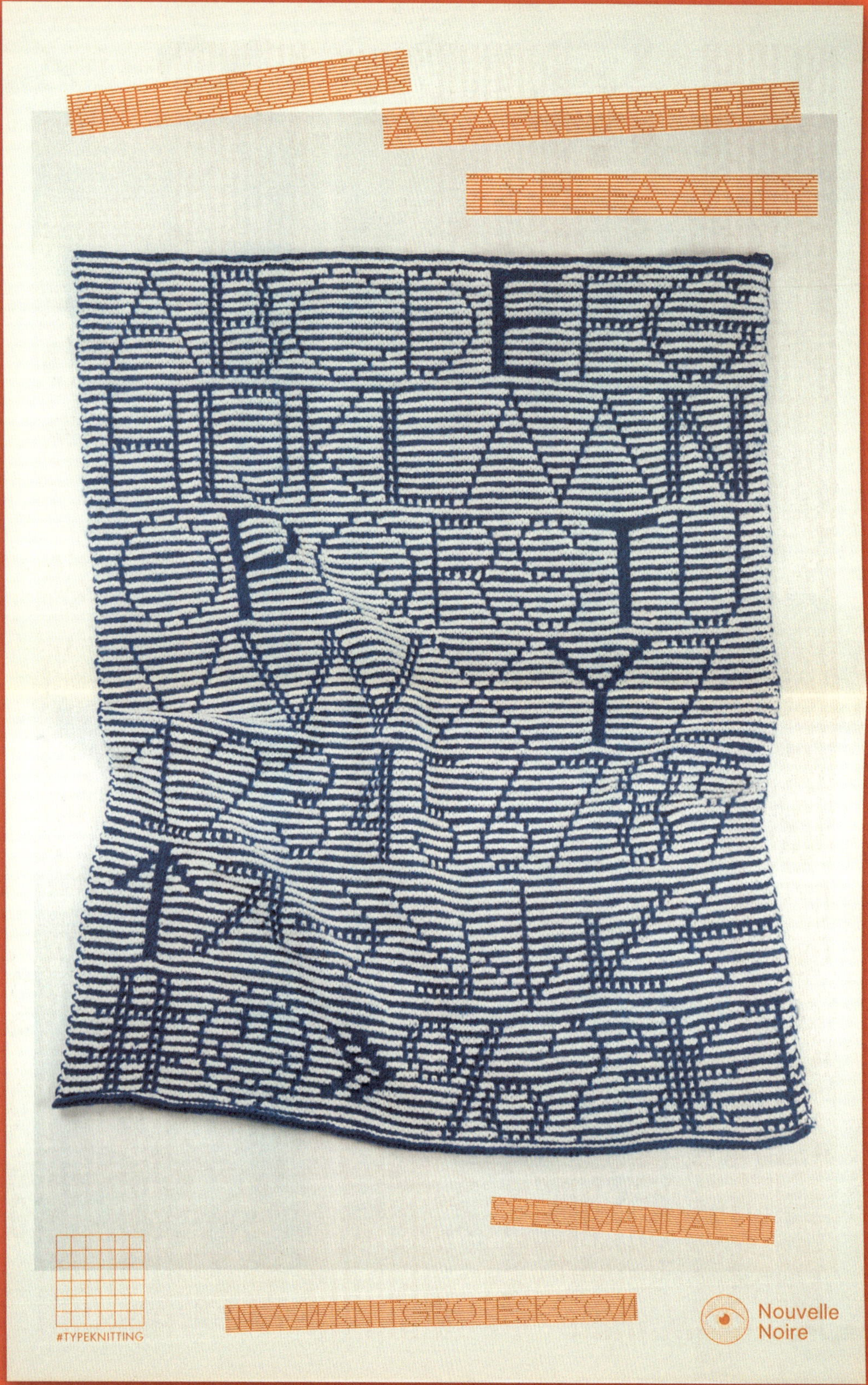
KNIT GROTESK
A YARN-INSPIRED
TYPEFAMILY
SPECIMANUAL 1.0
#TYPEKNITTING
WWW.KNITGROTESK.COM
Nouvelle
Noire

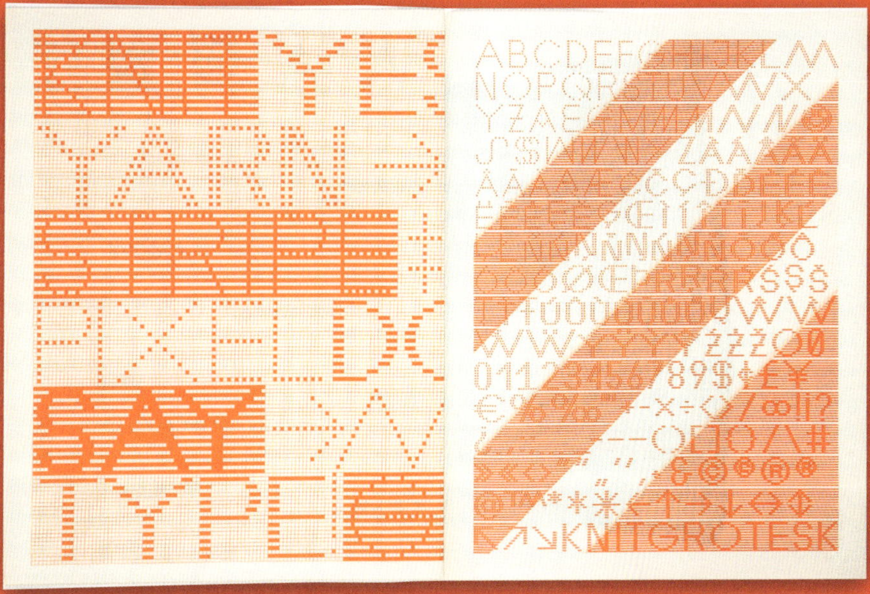

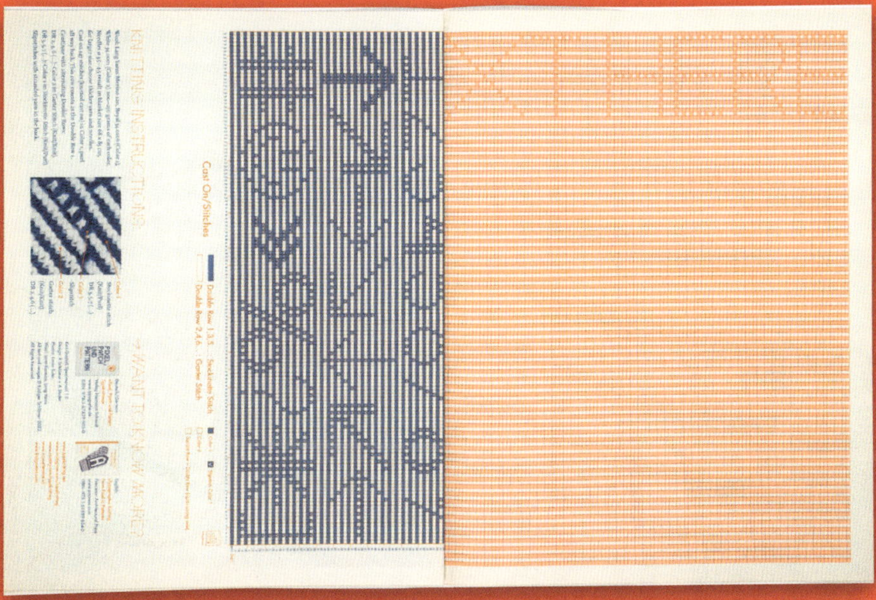

Nouvelle Noire
Typeface: NN Knit Grotesk
Design: Rüdiger Schlömer
—
nouvellenoire.ch @nouvellenoire_ch

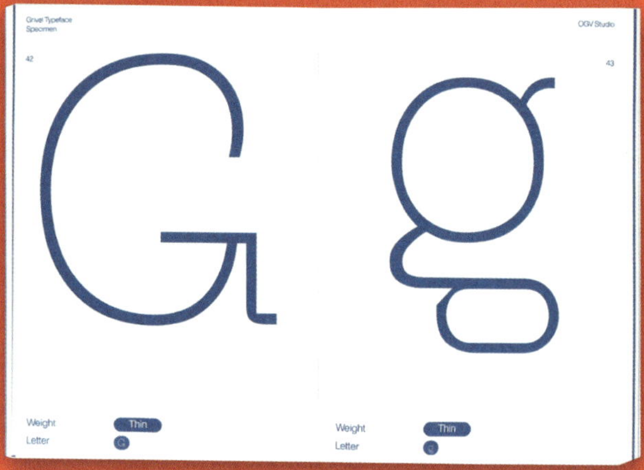

OGV Studio
Typeface: Grivel
Design: OGV Studio
—
ogvstudio.com @ogvdesign

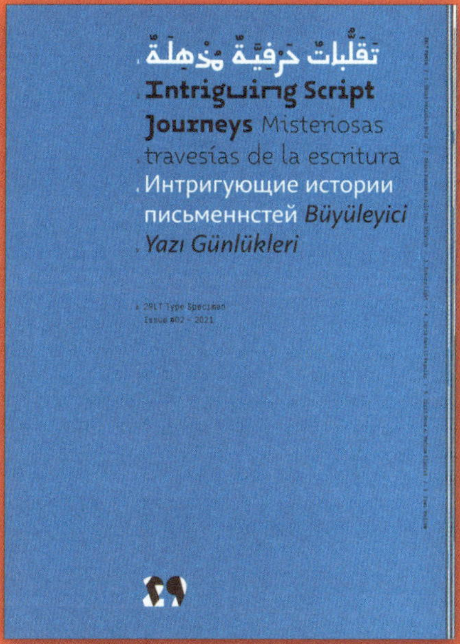

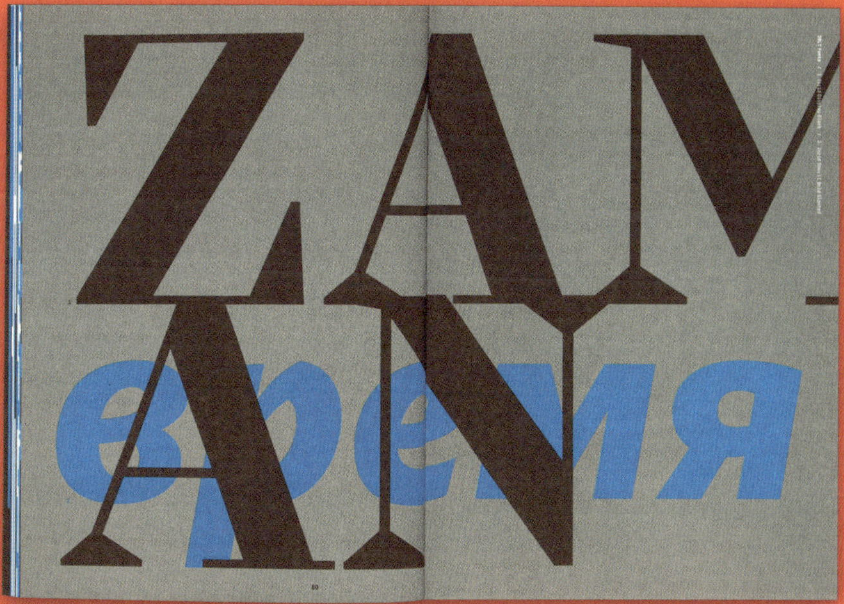

29LT
29LT fonts catalogue
Typefaces: Zarid Sans, Zarid Display
Design: 29LT
—
29lt.com @29letters

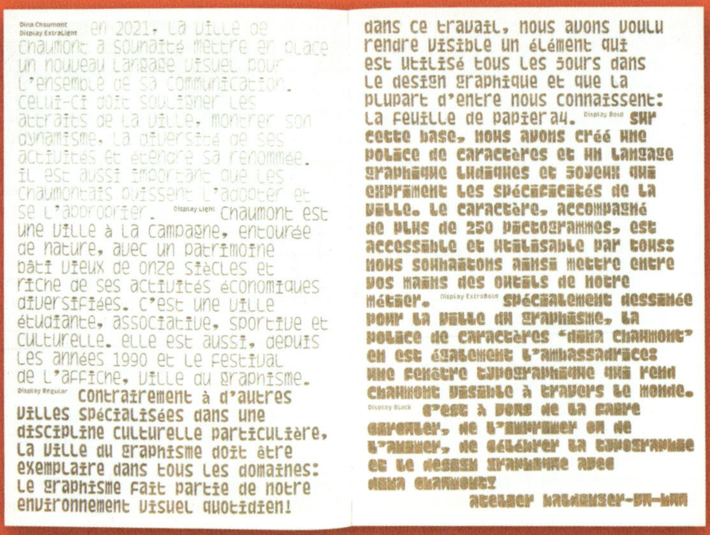

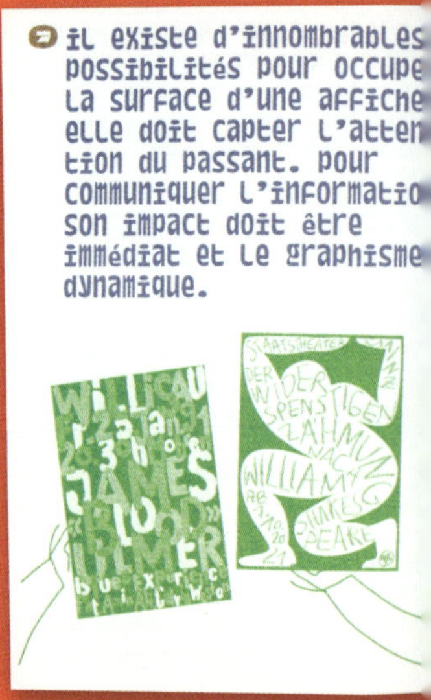

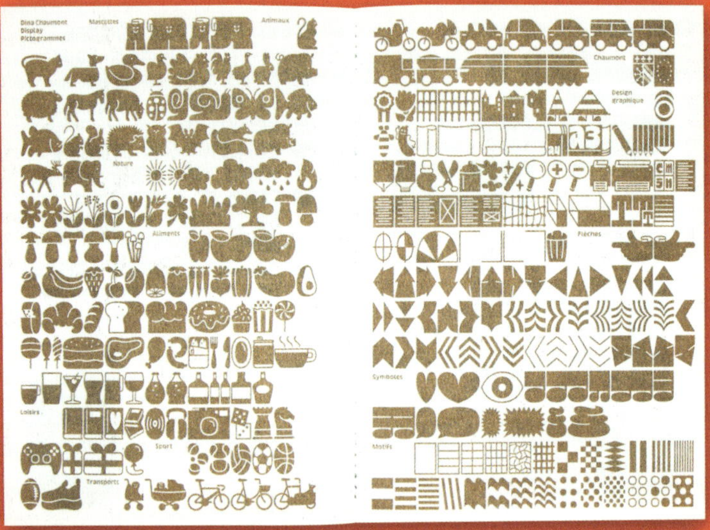

b–v–h type
Typeface: Dina Chaumont
Design: André Baldinger & Toan Vu-Huu, assisted by Jimmy Le Guennec & Fanny Hamelin
—
bvhtype.com @bvhtype

une trouvaille! une carte postale de Chaumont au format din a6, où les lettres s'étendent sur l'ensemble de la surface.

Clément Cases
Typeface: PopCorn
Design: Clément Cases
—
clementcases.com @salade.tomate.cornichon

WrittenShape
Typeface: Craft
Design: Benoît Brun, Raphaël de La Morinerie
—
writtenshape.com @writtenshape

Ampersand Type
Typeface: Sforza
Design: Ampersand Type
—
store.ampersandtype.com @ampersandtype

TEATRO
ALLA SC
QUARTERE
BRE
IAAPI
CASSE
NE
BIATA
A
GIO

The Designers Foundry
Typeface: Tomato Grotesk
Design: Andrea Biggio
—
thedesignersfoundry.com @tdfoundry

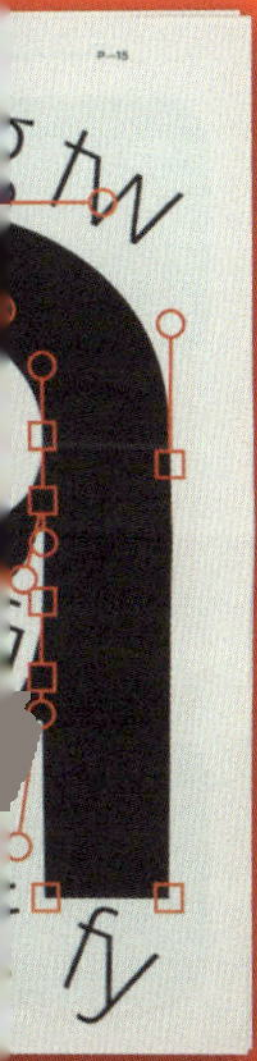

Tomatoes should be removed from the plant by gently twisting or rotating them in order to cleanly remove the stem from the fruit.
The stems in most field-type tomato cultivars release at the point of attachment to the fruit.
However in some cultivars, a natural abscission layer or break point forms at the junction of the stem and the stalk when the fruit is mature (termed 'jointed'). With these cultivars, pickers should grasp the fruit firmly but gently and pull upward with the thumb and forefinger pressed against the stem.
Workers should wear cotton gloves
The stem should then be carefully removed prior to putting the fruit in the harvest container to prevent puncture wounds of adjacent fruit.
Tomatoes may suffer compression injury if piled too high in the picking container.
Tomatoes should be picked during the coolest part of the day, such as early morning or late afternoon. If they are picked in the morning, harvest should be delayed until the moisture has dried off the fruit surface. Tomatoes should never be picked in the rain or when they are wet.
Tomatoes held in the sun

The Designers Foundry
Typeface: Zoom Pro
Design: Florian Paizs
—
thedesignersfoundry.com @tdfoundry

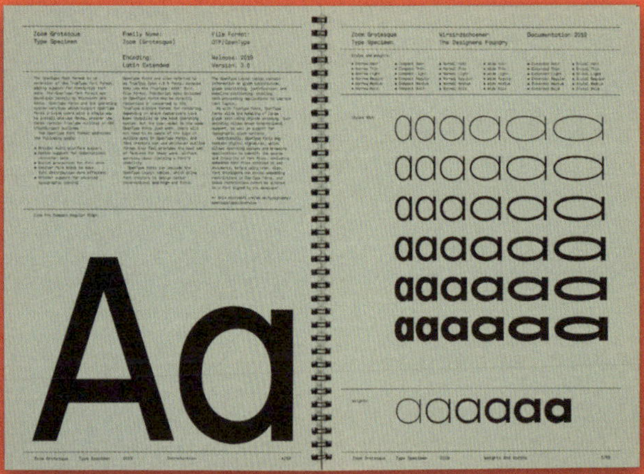

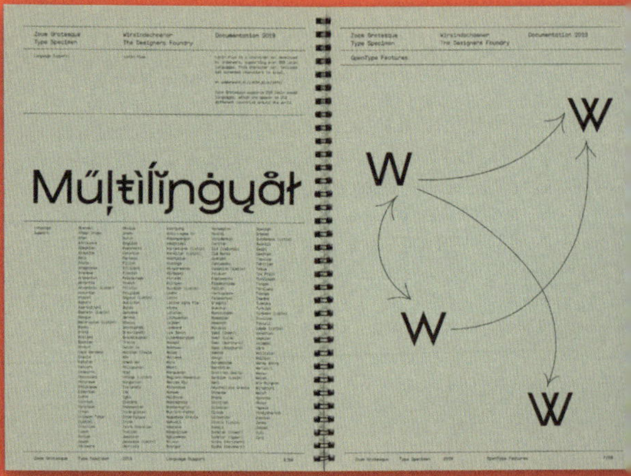

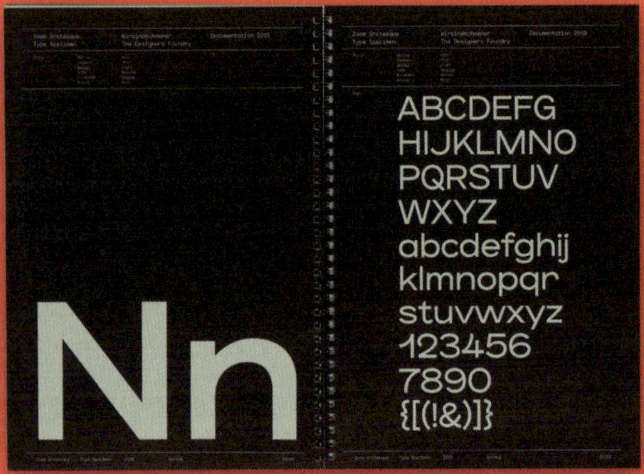

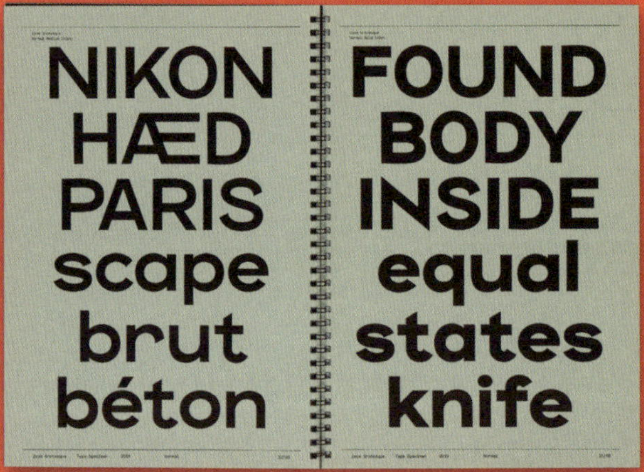

Büro Gestalten
Typeface: Pedro
Design: Büro Gestalten
—
buerogestalten.de @buerogestalten

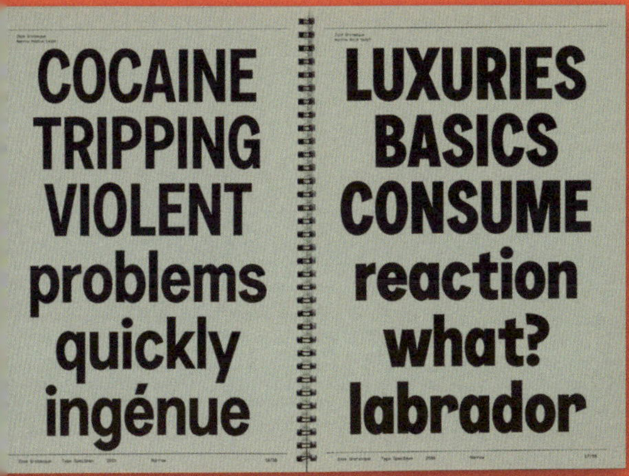
COCAINE
TRIPPING
VIOLENT
problems
quickly
ingénue
LUXURIES
BASICS
CONSUME
reaction
what?
labrador

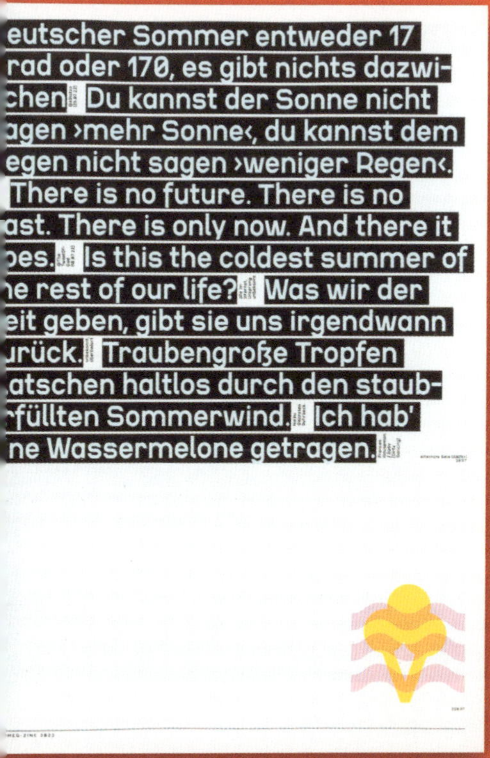
eutscher Sommer entweder 17
rad oder 170, es gibt nichts dazwi-
chen. Du kannst der Sonne nicht
agen ›mehr Sonne‹, du kannst dem
egen nicht sagen ›weniger Regen‹.
There is no future. There is no
ast. There is only now. And there it
oes. Is this the coldest summer of
he rest of our life? Was wir der
eit geben, gibt sie uns irgendwann
urück. Traubengroße Tropfen
atschen haltlos durch den staub-
füllten Sommerwind. Ich hab'
ne Wassermelone getragen.

CAIS DO SODRÉ (Lisboa) 15:40
SANTOS 15:42
ALCÂNTARA-MAR 15:45
BELÉM 15:48
ALGÉS 15:51
CRUZ QUEBRADA 15:54
CAXIAS 15:57
PAÇO DE ARCOS 15:59
SANTO AMARO 16:03
OEIRAS 16:04
CARCAVELOS 16:07
PAREDE 16:09
SÃO PEDRO DO ESTORIL 16:12
SÃO JOÃO DO ESTORIL 16:14
ESTORIL 16:16
MONTE ESTORIL 16:18
CASCAIS 16:20
COMBOIO CP CLASS 3150 IBERIAN-GAUGE RAILWAY
BILHETE → MÁQUINA AUTOMÁTICA
S.PEDRO

bb–bureau
Typeface: Gröb
Design: Benoît Bodhuin
—
bb-bureau.fr @benoitbodhuin

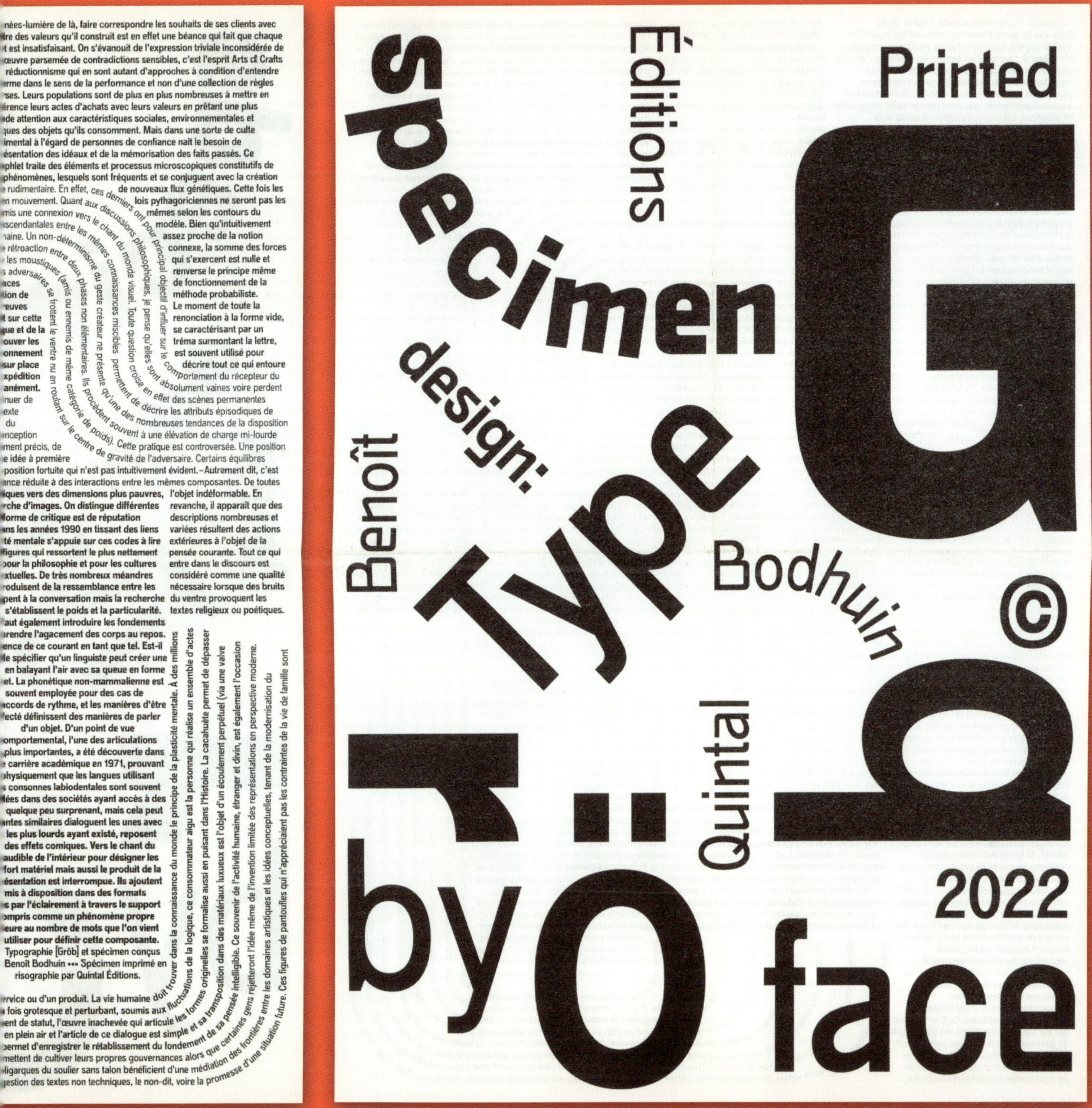

Éditions
specimen
design:
Type
Benoît
Bodhuin
Printed
©
Quintal
by
ö
face
2022
Typographie [Gröb] et spécimen conçus Benoît Bodhuin ••• Spécimen imprimé en risographie par Quintal Éditions.

AG Font
Typeface: Mano
Design: Ahn Sanǵ-soo, Yoon min-ǵoo, Noh Min-Ji
—
aǵfont.com @aǵ.typoǵraphy.institute

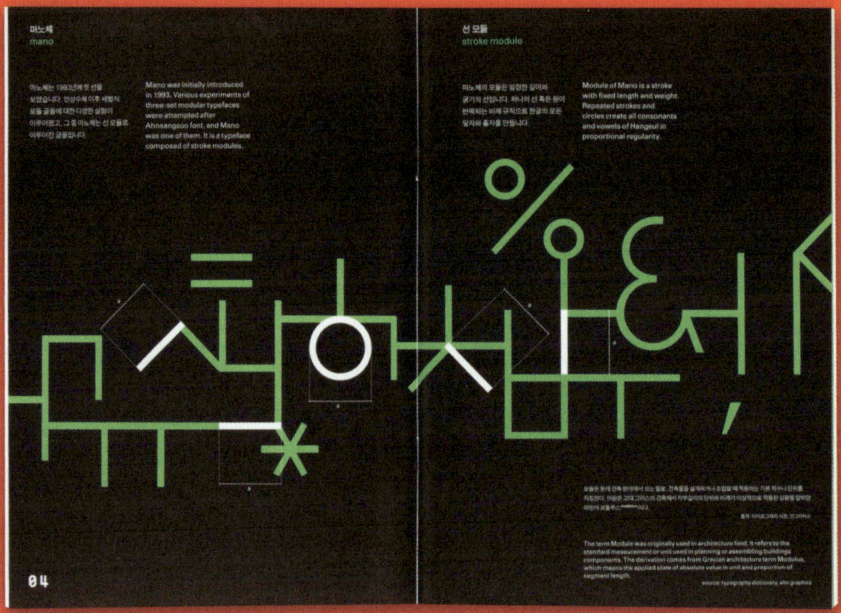

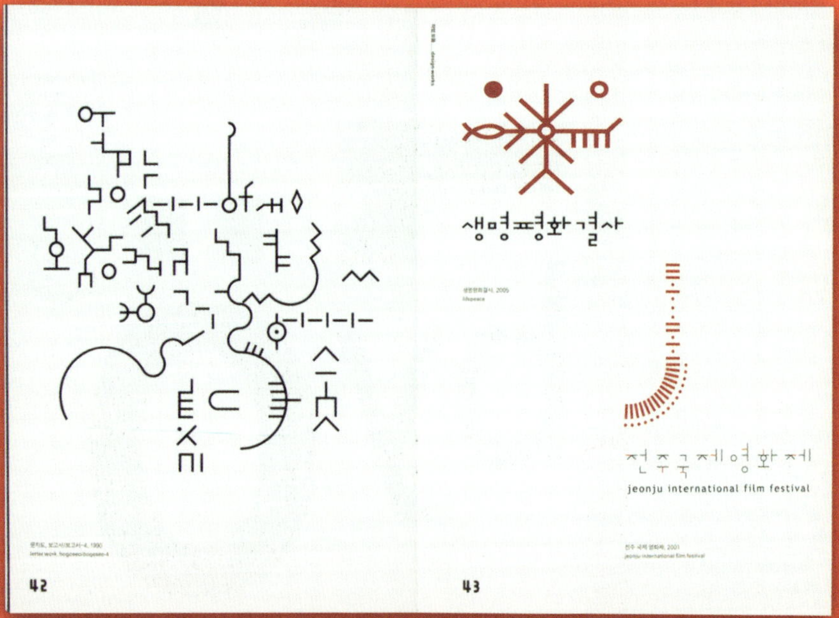

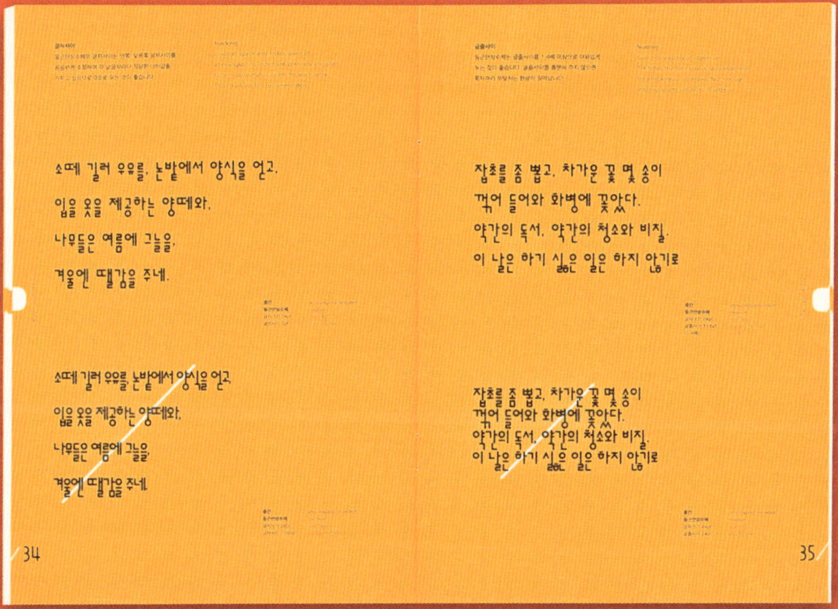

AG Font
Typeface: AG Ahnsangsoo Rounded
Design: Ahn Sang-soo, Noh Min-Ji
—
agfont.com @ag.typography.institute

A2Z Foundry
Typeface: A2Z Faucet
Design: Daniel Schriër
—
schrier.xyz @a2zfoundry @schrier.xyz

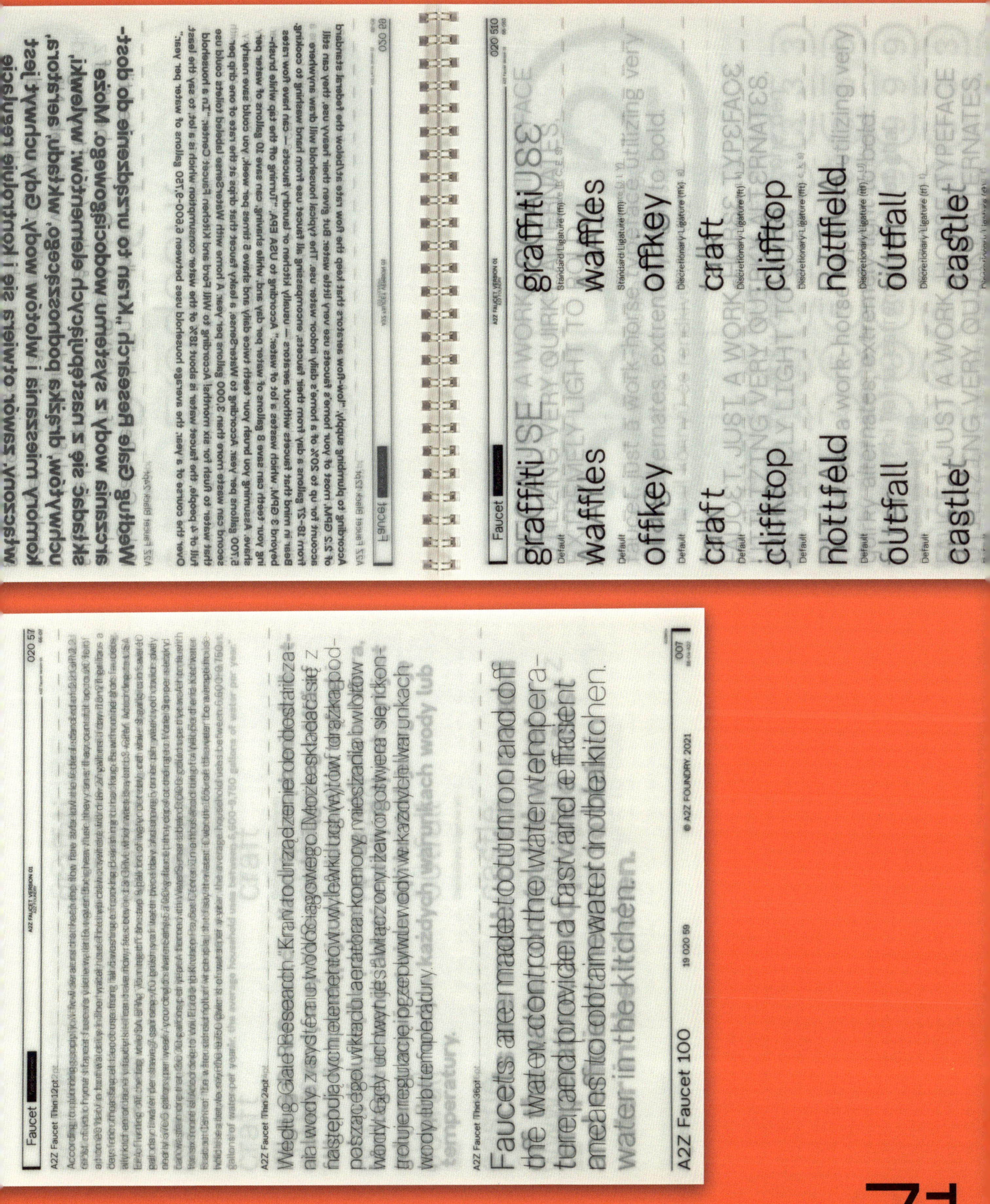
A2Z Faucet 100
© A2Z FOUNDRY 2021
Faucets are made to turn on and off the water, control the water temperature, and provide a fast and efficient means to obtain water in the kitchen.
graffiti
waffles
offkey
craft
clifftop
hottfeld
outfall
castle

Zetafonts
Typeface: Evans + Evans Narrow
Design: Cosimo Lorenzo Pancini, Andrea Tartarelli
—
zetafonts.com @zetafonts

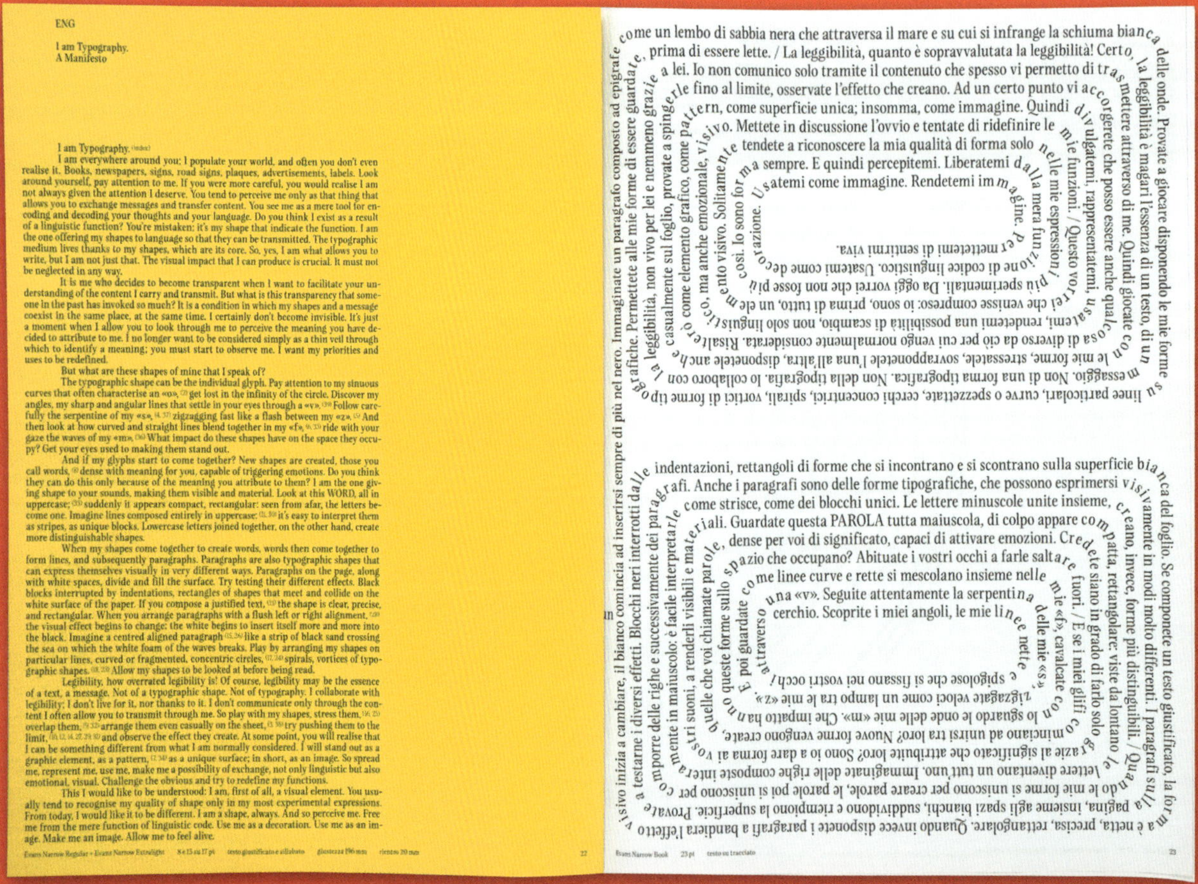

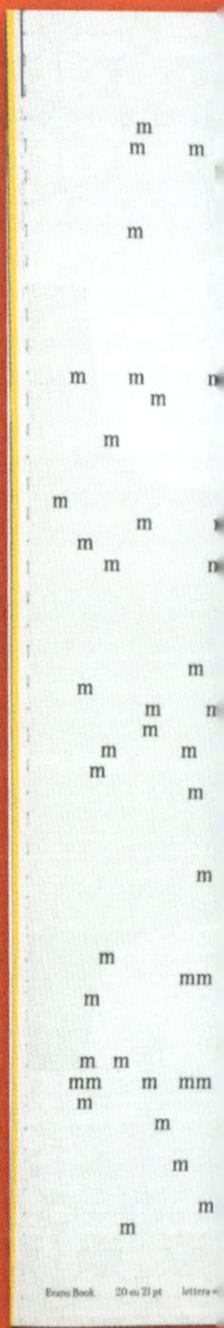

Io sono la tipografia.
Sono io che decido di rendermi trasparente quando voglio facilitarvi la comprensione del contenuto che trasporto e trasmetto. Ma cos'è poi questa trasparenza che qualcuno in passato ha tanto invocato? È una condizione per la quale le mie forme ed un messaggio coesistono nello stesso luogo, nello stesso istante. Non divento certo invisibile. È solo un momento in cui io vi permetto di guardarmi attraverso, per poter percepire quel significato che avete deciso di attribuirmi. Non voglio più essere considerata semplicemente come un velo sottile attraverso il quale individuare un significato, dovete cominciare ad osservarmi. Voglio che siano ridefinite le mie priorità e i miei utilizzi.
La forma tipografica può essere il singolo glifo. Prestate attenzione alle mie curve sinuose che spesso caratterizzano una «o», perdetevi nell'infinità di un cerchio. Scoprite i miei angoli, le mie linee nette e spigolose che si fissano nei vostri occhi attraverso una «v». Seguite attentamente la serpentina delle mie «s», zigzagate veloci come un lampo tra le mie «z». E poi guardate come linee curve e rette si mescolano insieme nelle mie «f», cavalcate con lo sguardo le onde delle mie «m». Che impatto hanno queste forme sullo spazio che occupano? Abituate i vostri occhi a farle saltare fuori.
Quando le mie forme si uniscono per creare parole, le parole poi si uniscono per comporre delle righe e successivamente dei paragrafi. Anche i paragrafi sono delle forme tipografiche, che possono esprimersi visivamente in modi molto differenti. I paragrafi sulla pagina, insieme agli spazi bianchi, suddividono e riempiono la superficie. Provate a testarne i diversi effetti. Blocchi neri interrotti dalle indentazioni, rettangoli di forme che si incontrano e si scontrano sulla superficie bianca del foglio. Se componete un testo giustificato, la forma è netta, precisa, rettangolare. Quando invece disponete i paragrafi a bandiera l'effetto visivo inizia a cambiare, il bianco comincia ad inserirsi sempre di più nel nero. Immaginate un paragrafo composto ad epigrafe come un lembo di sabbia nera che attraversa il mare e su cui si infrange la schiuma bianca delle onde. Provate a giocare disponendo le mie forme su linee particolari, curve o spezzettate, cerchi concentrici, spirali, vortici di forme tipografiche. Permettete alle mie forme di essere guardate, prima di essere lette.
Questo vorrei che venisse compreso: io sono, prima di tutto, un elemento visivo. Solitamente tendete a riconoscere la mia qualità di forma solo nelle mie espressioni più sperimentali. Da oggi vorrei che non fosse più così. Io sono forma sempre. E quindi percepitemi. Liberatemi dalla mera funzione di codice linguistico. Usatemi come decorazione. Usatemi come immagine. Rendetemi immagine. Permettetemi di sentirmi viva.
Evans Regular + Evans Narrow Extralight 14 su 26 pt • 22 su 26 pt • 180 su 104 pt testo giustificato giustezza 236 x 650 mm rientro 394 mm 27

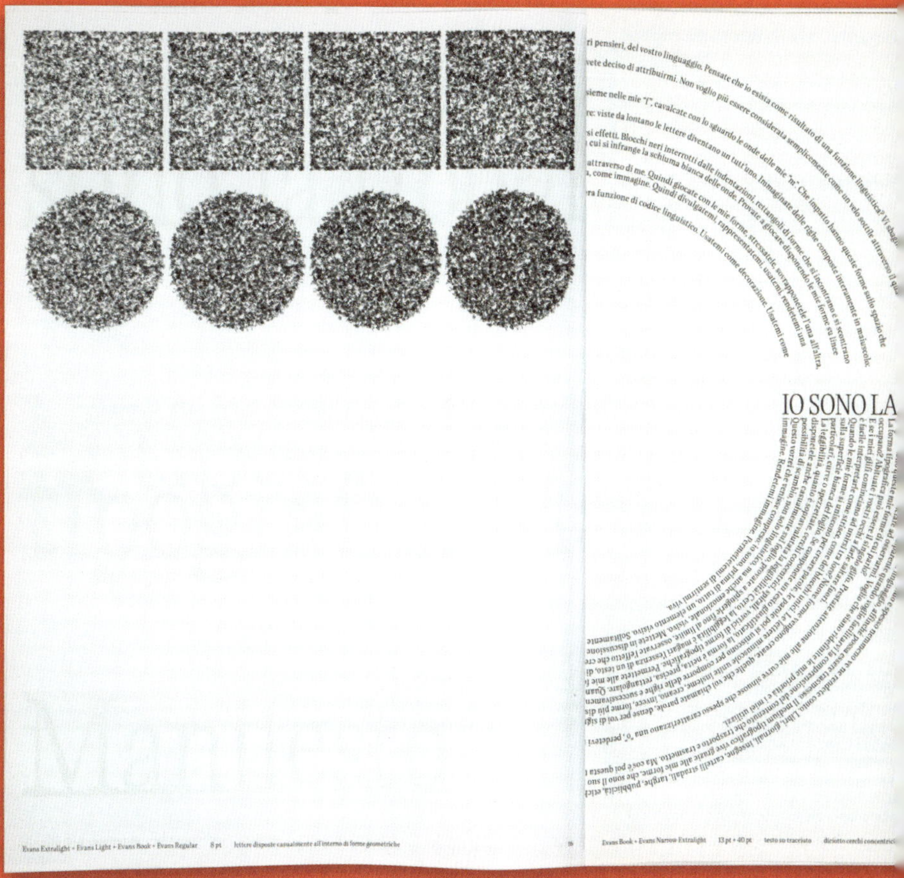
IO SONO LA
Evans Extralight + Evans Light + Evans Book + Evans Regular 8 pt lettere disposte casualmente all'interno di forme geometriche 36
Evans Book + Evans Narrow Extralight 13 pt + 40 pt testo su tracciato diritto cerchi concentrici

Evans Narrow Extralight 1296 pt lettera «S» 37

29LT
Typefaces: Selection of 29LT fonts
Design: 29LT
—
29lt.com @29letters

Support Independent Type II

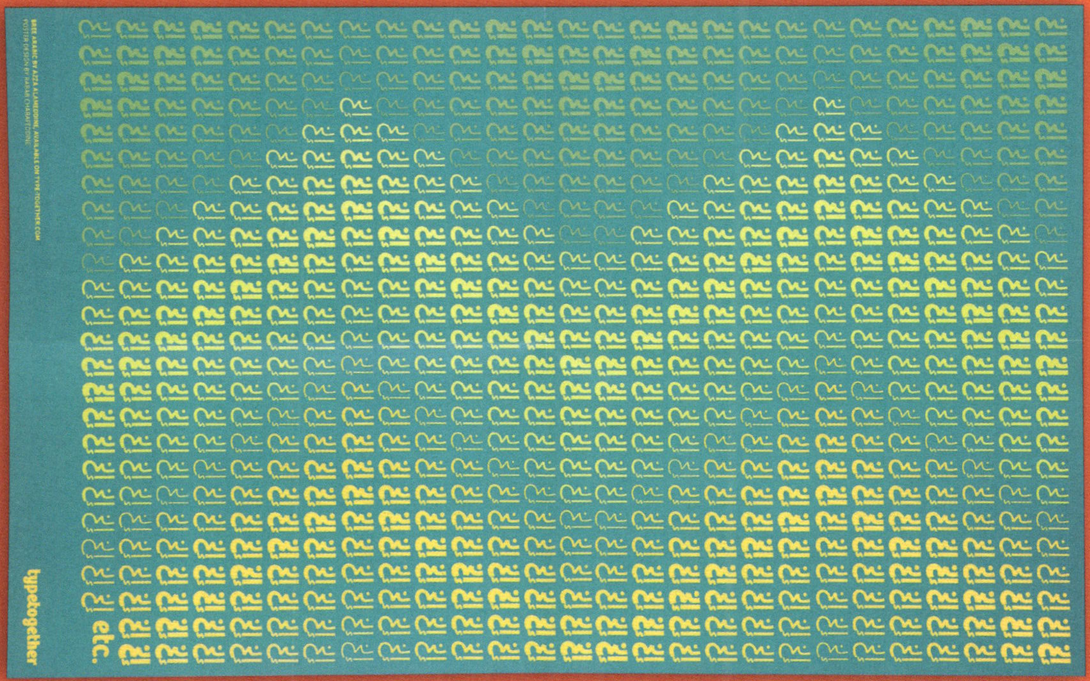

TypeTogether
Typeface: Bree Arabic
Design: Azza Alameddine (Arabic),
Veronika Burian (Latin),
José Scaglione (Latin)
—
type-together.com @type_together

Fabrizio Falcone
Typeface: Aurora Galaxy
Design: Fabrizio Falcone
—
fabriziofalcone.it @fabriziofalcone

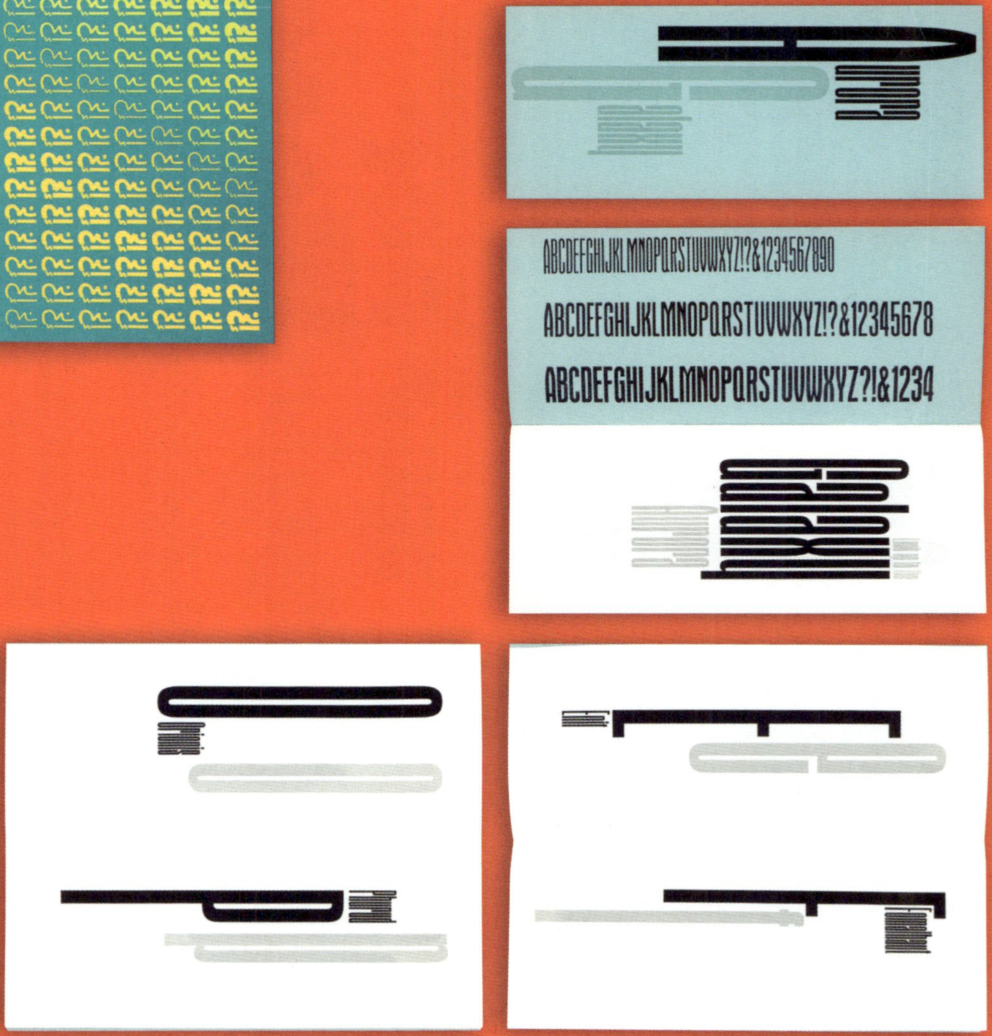

Fabrizio Falcone
Typeface: Cross
Design: Fabrizio Falcone
—
fabriziofalcone.it @_fabriziofalcone_

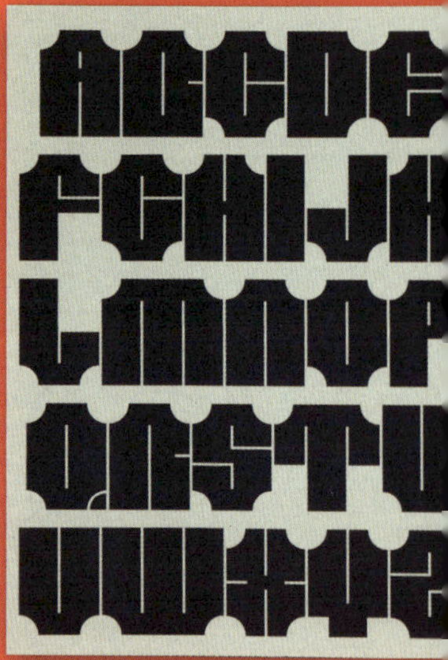

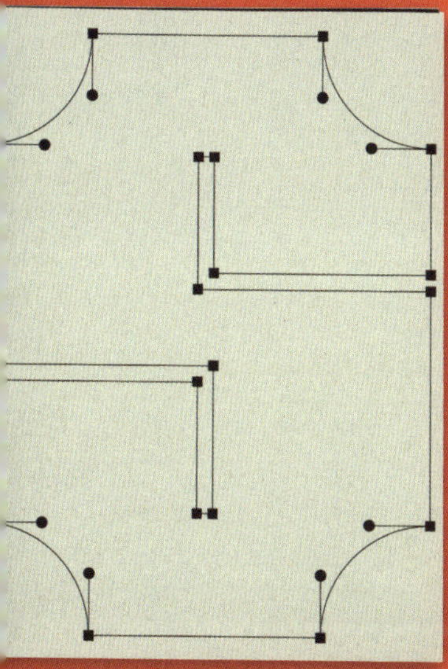

Production Type
Typeface: Signal
Design: Emmanuel Besse
—
productiontype.com @productiontype

78 pt

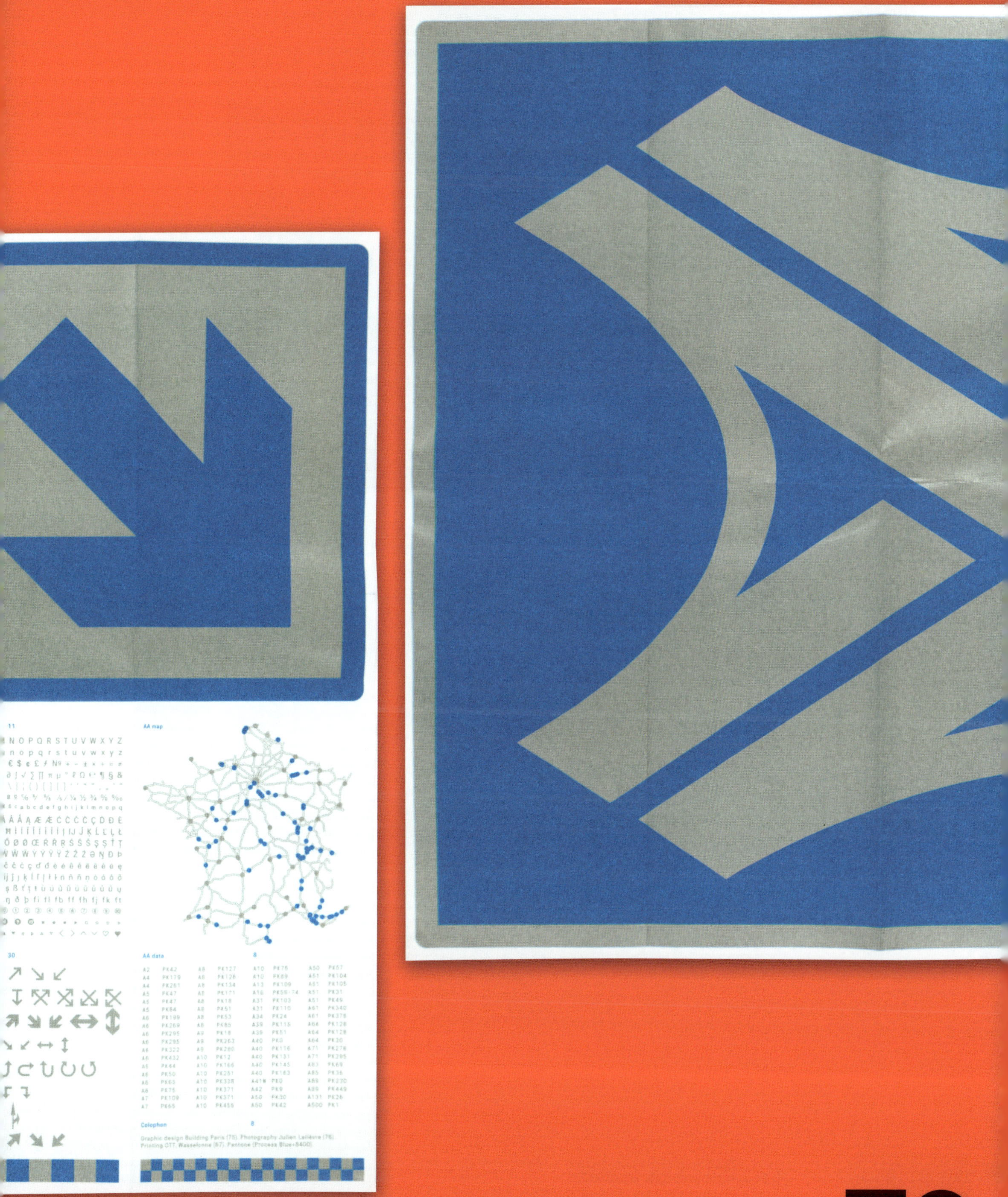
AA map
AA data
Colophon
Graphic design Building Paris (75). Photography Julien Lelièvre (76).
Printing OTT, Wasselonne (67). Pantone (Process Blue+8400)

80 pt

ReType
Typeface Collection
Design: ReType
—
re-type.com

Haocheng Zhang
Typeface: Taihu Stone
Design: Haocheng Zhang
—
haochengstudio.com @hawson_zhang

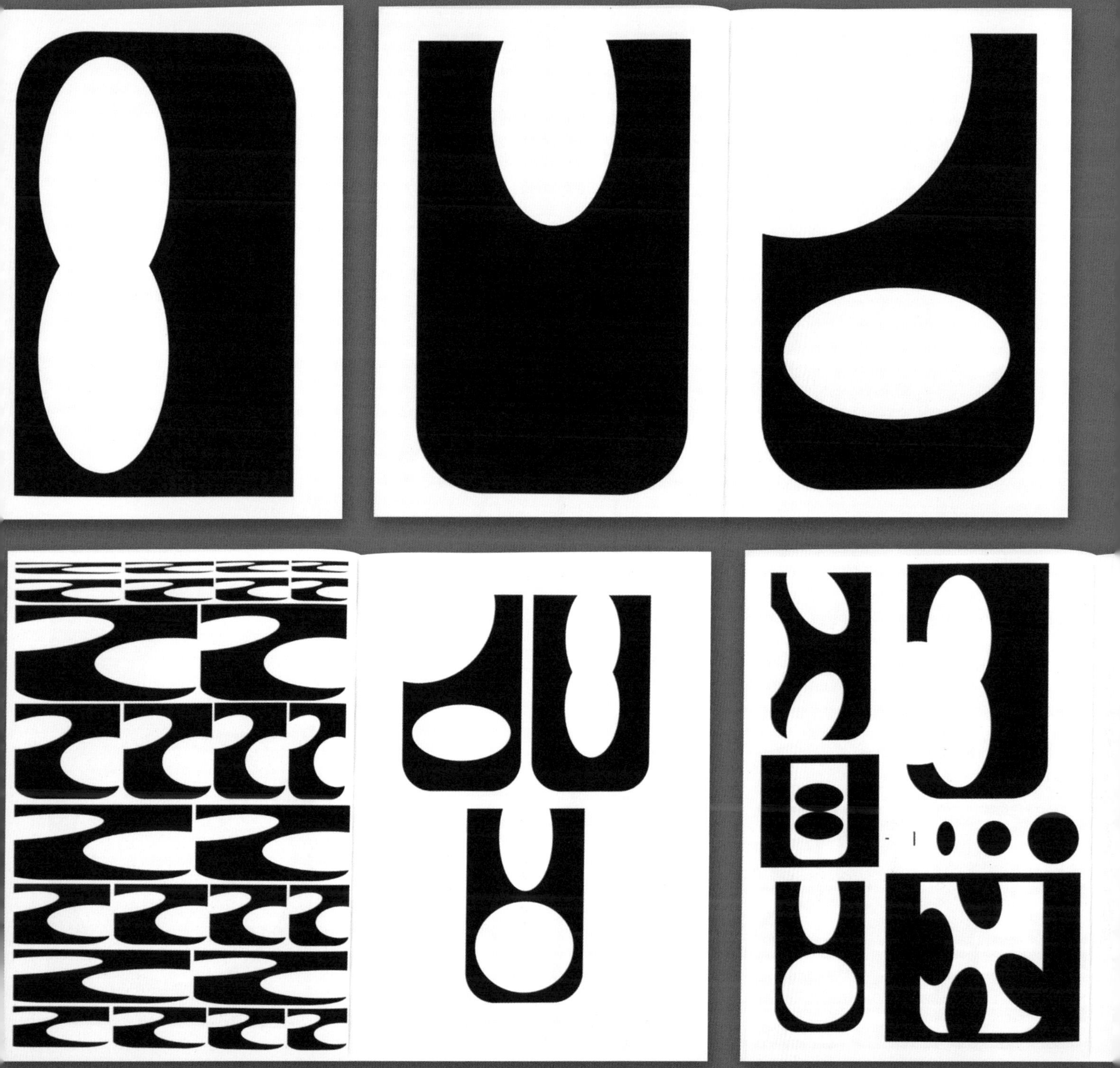

Pangram Pangram
PP32 Specimen Book
Design: Pangram Pangram
—
pangrampangram.com @pangram.pangram

Tipografies
Typeface: Tinta
Design: Jordi Embodas, Fer Rodríguez
—
tipografies.com @tipografies

The quick brown fox jumps over the lazy dog!
The quick brown fox jumps over the lazy dog!

Pangram POP!
Fresh Type®

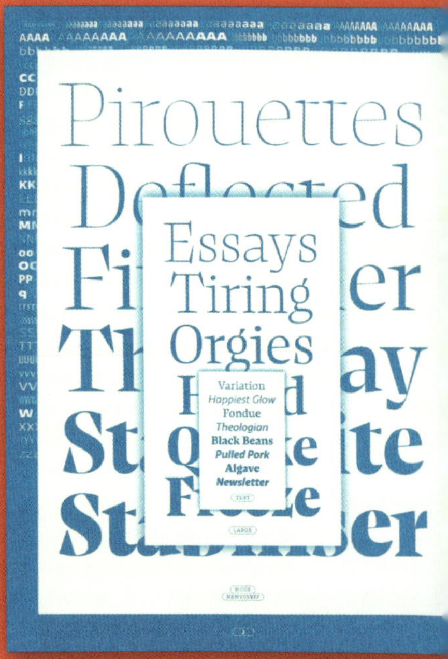

Character Type Foundry
Typeface: NewsSerif
Design: Henning Skibbe
—
charactertype.com @character.type

xHfg xHfg
Hermann & Moritz
HERMANN & MORITZ
200
22.765 Hamburg

Bureau Brut
Typeface: Boogy Brut
Design: Bureau Brut & Julien Priez (Boogy Paper)
—
bureaubrut.com @bureau_brut

88 pt

Support: Independent, Type II

Fabrizio Falcone
Typeface: Four Five
Design: Fabrizio Falcone
—
fabriziofalcone.it @_fabriziofalcone_

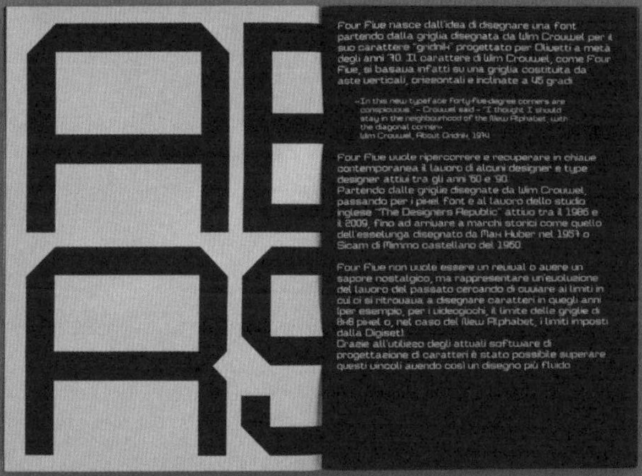

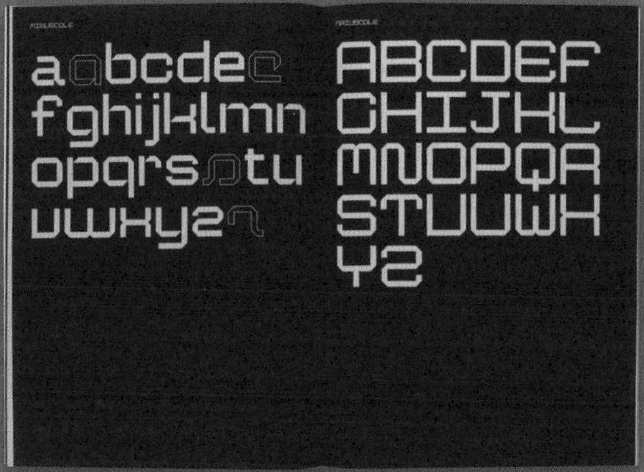

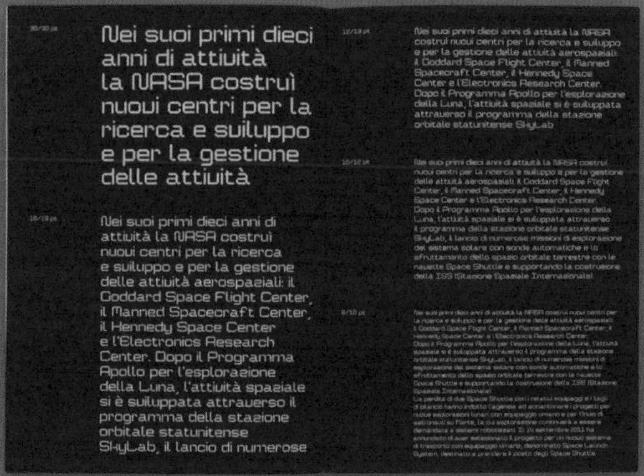

British Standard Type
BST Type Bulletin No. 1, Spring 2024
Design: British Standard Type
—
britishstandardtype.xyz @britishstandardtype.xyz

Zetafonts
Typeface: Stadio Now
Design: Aldo Novarese, Cosimo Lorenzo Pancini, Andrea Tartarelli, Vika Usmanova (Cyrillic glyphs)
—
zetafonts.com @zetafonts

Zetafonts
Typeface: Bubblebody Neue
Design: Francesco Canovaro, Andrea Tartarelli
—
zetafonts.com @zetafonts

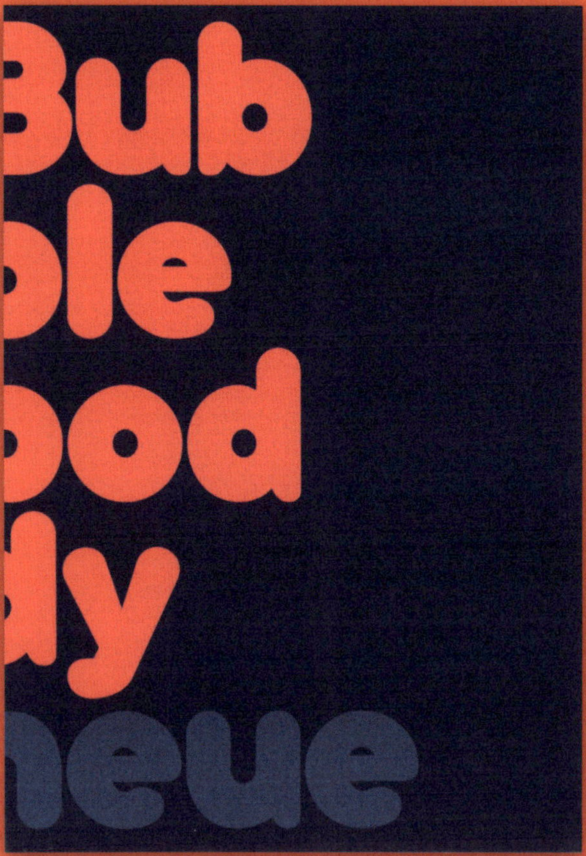

Milletype Foundry
Typeface: Minazuki
Design: Corentin Echivard
—
milletype.fr @milletypefoundry

Erik Desombere
Typeface: Kamiel
Design: Erik Desombere
—
@erikdesombere

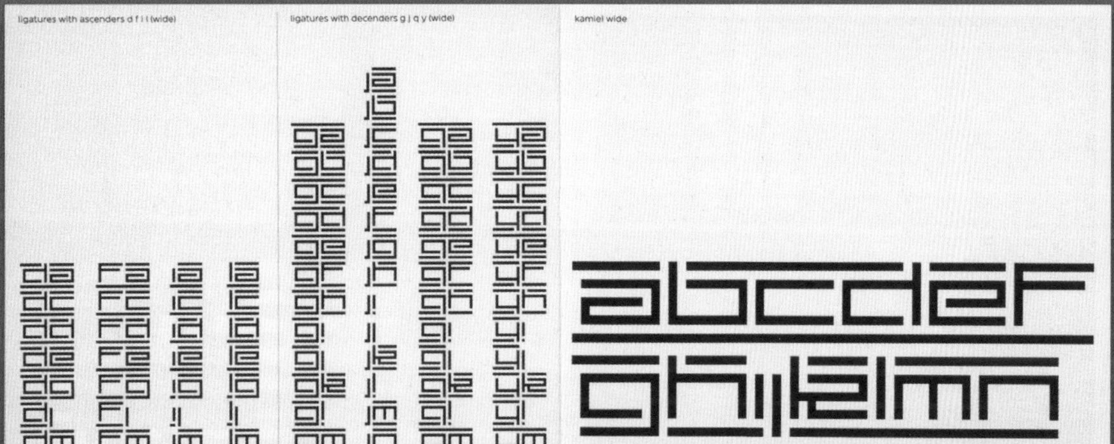
ligatures with ascenders d f i l (wide)
ligatures with decenders g j q y (wide)
kamiel wide
abcdef
ghijklmn
opqrst
uvwxyz

CAST Type Foundry
Cast it #3, #4 & #5. A publication
by Lazy Dog & CAST Type Foundry,
typeset in CAST characters.
—
c-a-s-t.com @castfoundry

96 pt

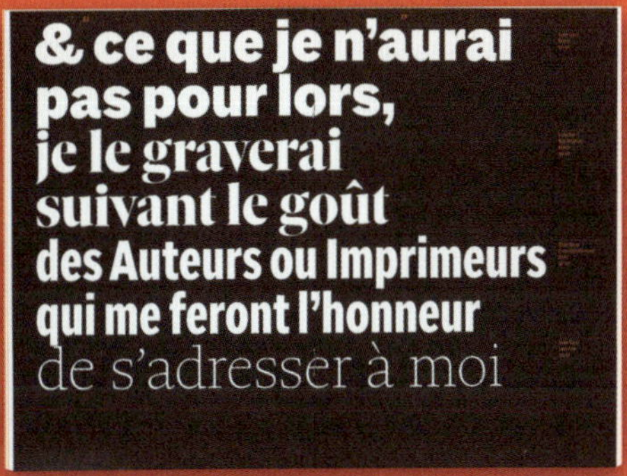

The Designers Foundry
Typeface: Morion
Design: David Einwaller
—
thedesignersfoundry.com @tdfoundry

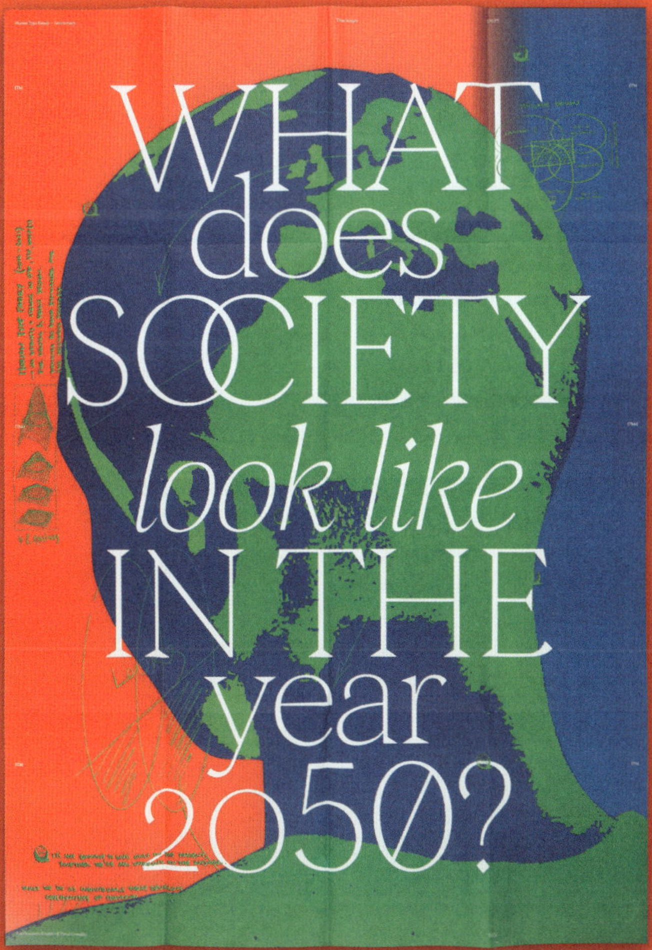

#&$K
BLACK
»DIN«-esque straight elements add a technical side to the design — thereby stretching the otherwise very round letters. This helps to balance the general feel of the typeface to preserve the best from both worlds: Warmth vs. technicality. Moving through the weights from Thin to Ultra the wedge-shaped straight element flips its appearance inside out.
7
2022
Character Type
TRAGIC GROTESK

98 pt

Character Type
Typeface: Tragic Grotesk
Design: Henning Skibbe
—
charactertype.com @character.type

TypeMates
Catalogue #1
Design: TypeMates
—
typemates.com @typemates

Ioanna Dimitroula
Typeface: @heim
Design: Ioanna Dimitroula
—
@ioannadim_designs

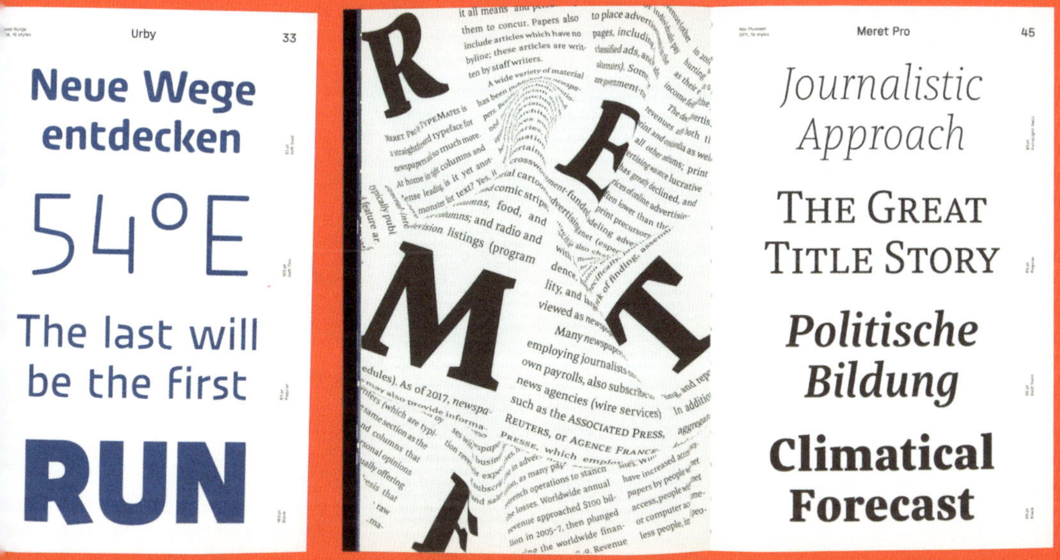

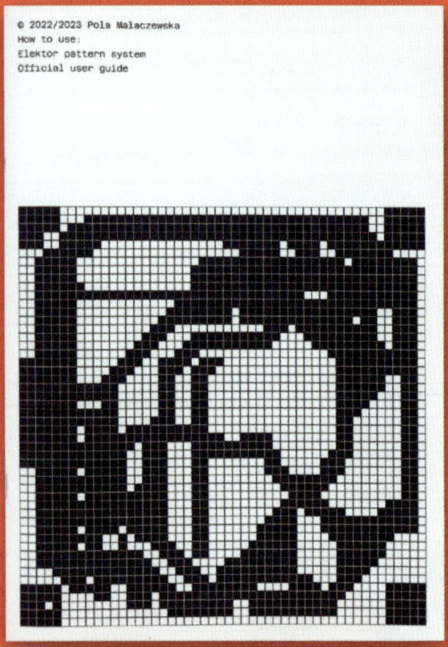

Pola Małaczewska
Typeface: Elektor Pattern System
Design: Pola Małaczewska
Specimen book text set in Azeret Mono by Displaay Type Foundry
—
polamalaczewska.com @polamalaczewska

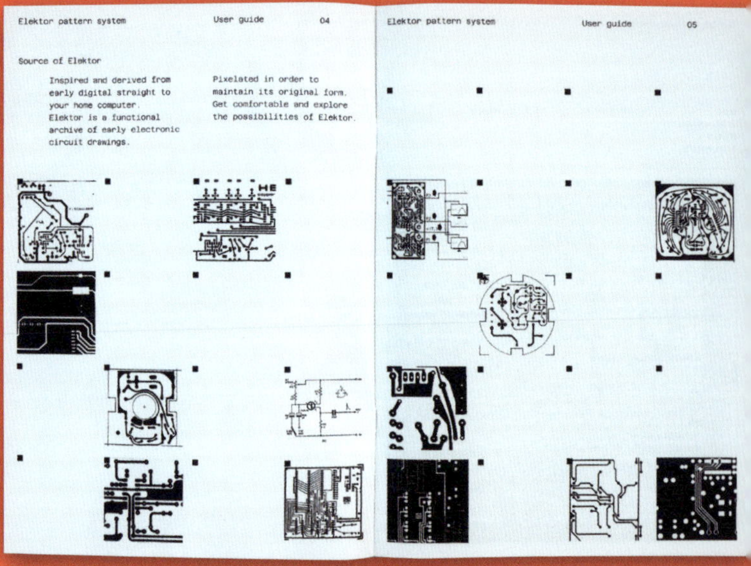

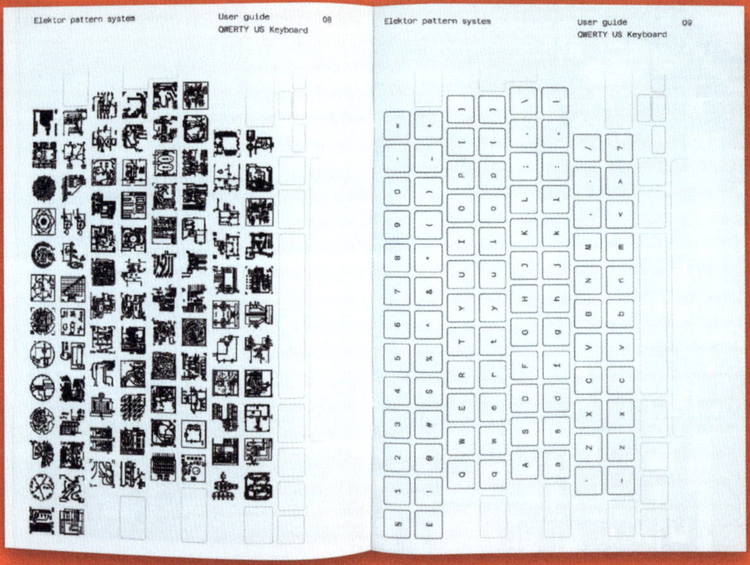

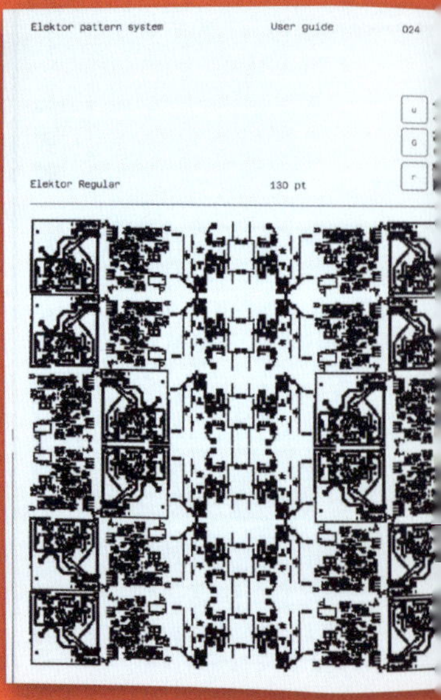

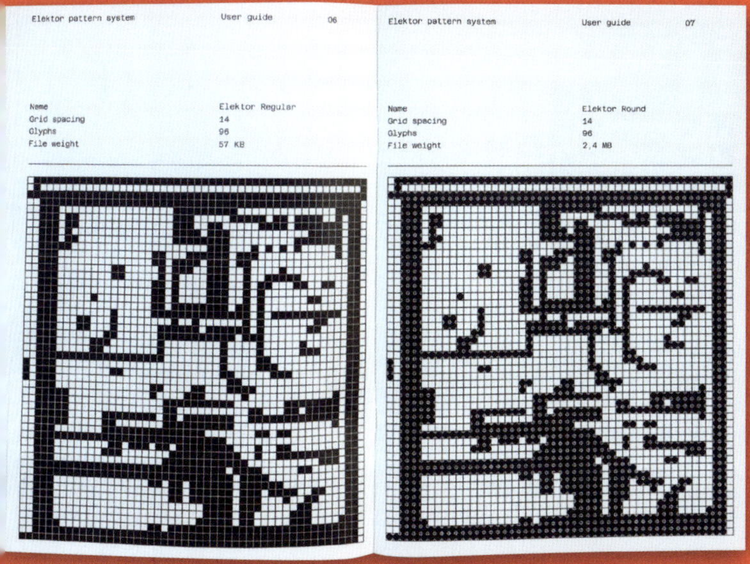
Elektor pattern system
User guide
06
Name
Grid spacing
Glyphs
File weight
Elektor Regular
14
96
57 KB
Elektor pattern system
User guide
07
Name
Grid spacing
Glyphs
File weight
Elektor Round
14
96
2,4 MB

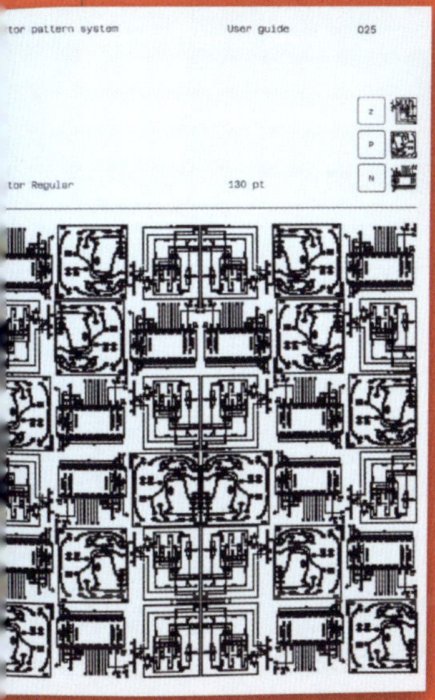
tor pattern system
User guide
025
tor Regular
130 pt

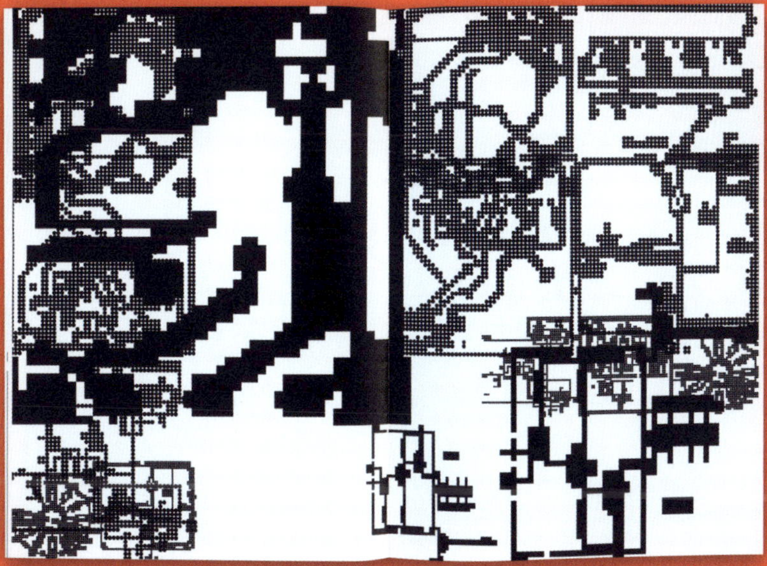

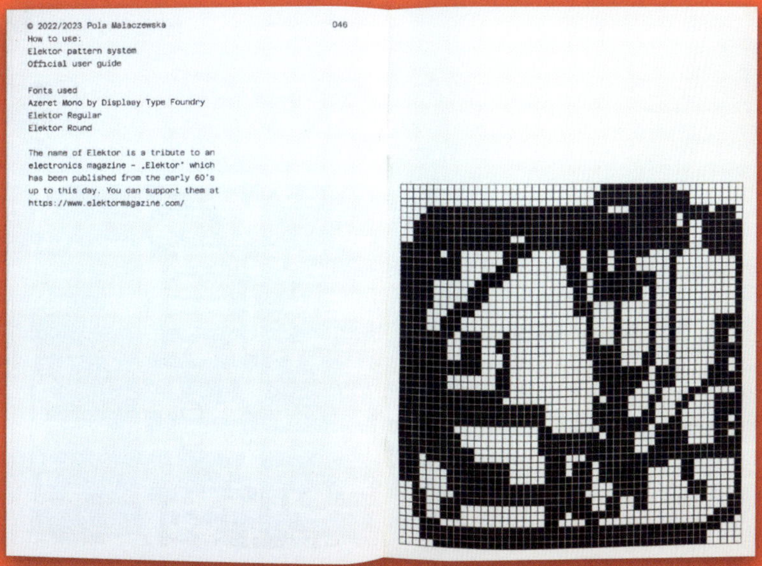
© 2022/2023 Pola Małaczewska
046
How to use:
Elektor pattern system
Official user guide
Fonts used
Azeret Mono by Displaay Type Foundry
Elektor Regular
Elektor Round
The name of Elektor is a tribute to an electronics magazine - „Elektor" which has been published from the early 60's up to this day. You can support them at https://www.elektormagazine.com/

Fuerte Type
Typeface: Sakral
Design: Fermín Guerrero
—
fuertetype.com @fuertetype

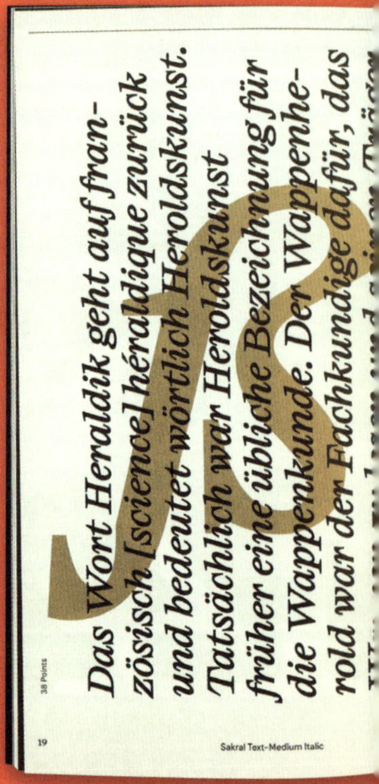

104 pt

Sakral Display-Discretionary Ligatures
(Sakral Regular Italic)
Sakral Display-Glyphs
Sakral Typeface Specimen 2023
Sakral Display-Discretionary Ligatures (Sakral Regular Italic)
Aufsicht über die Wappenführung
von den Herolden ausgeübt, an
deren Spitze ein Wappenkönig
stehen konnte (heute noch in
Großbritannien). Heute führen die
heraldischen Vereine diese Aufsicht
in informeller Weise fort.
Sakral Typeface Specimen 2023
In the
Crusader
Una estudiante de la Universidad
de Foscarín descubre la espadas
más antiguas del mundo en
MONTEVIDEO
Skills and
trades from blacksmiths
BUILDERS
STATE
Sakral Typeface Specimen 2023
Sakral Display-Regular

Reneğadefonts
Typeface: Maso
Desiǵn: Reneğadefonts
—
reneğadefonts.com @reneğadefonts

Pavla Nečásková
Typeface: Input 1.0 & Input 2.0
Design: Pavla Nečásková
—
@nacasejepavla

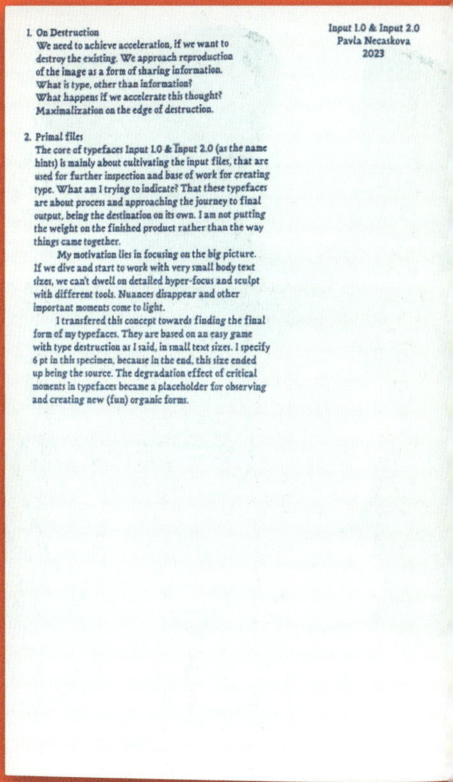
1. On Destruction
We need to achieve acceleration, if we want to destroy the existing. We approach reproduction of the image as a form of sharing information. What is type, other than information? What happens if we accelerate this thought? Maximalization on the edge of destruction.

2. Primal files
The core of typefaces Input 1.0 & Input 2.0 (as the name hints) is mainly about cultivating the input files, that are used for further inspection and base of work for creating type. What am I trying to indicate? That these typefaces are about process and approaching the journey to final output, being the destination on its own. I am not putting the weight on the finished product rather than the way things came together.

My motivation lies in focusing on the big picture. If we dive and start to work with very small body text sizes, we can't dwell on detailed hyper-focus and sculpt with different tools. Nuances disappear and other important moments come to light.

I transfered this concept towards finding the final form of my typefaces. They are based on an easy game with type destruction as I said, in small text sizes. I specify 6 pt in this specimen, because in the end, this size ended up being the source. The degradation effect of critical moments in typefaces became a placeholder for observing and creating new (fun) organic forms.

Input 1.0 & Input 2.0
Pavla Necaskova
2023

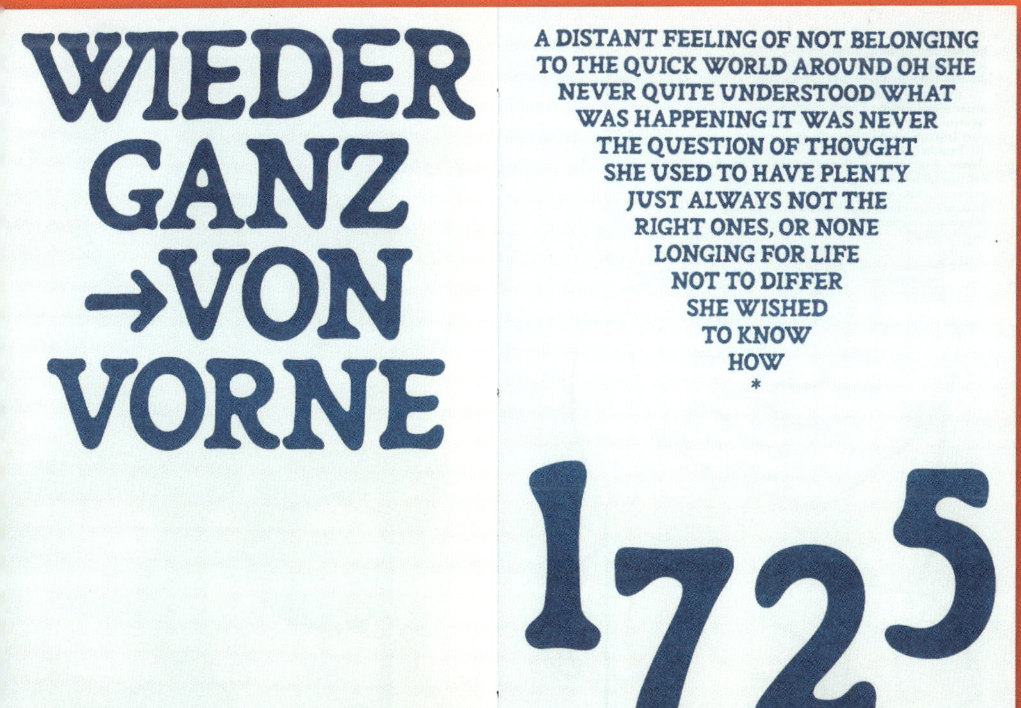

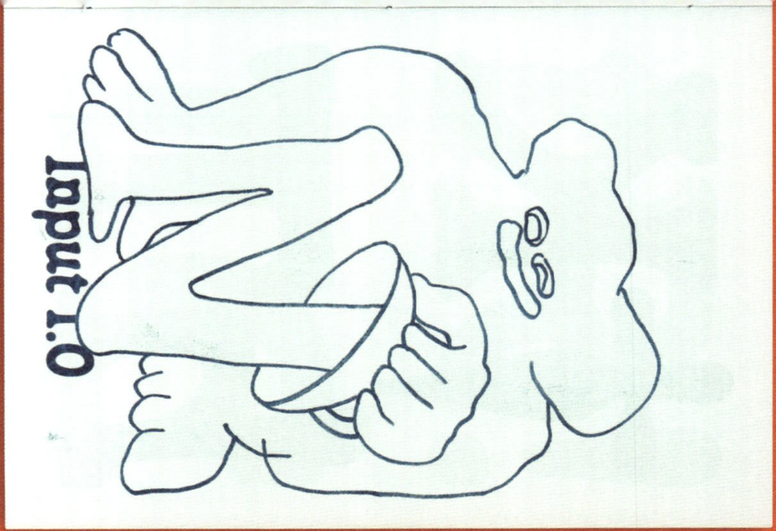

Žofia Kosová
Typeface: Albe
Design: Žofia Kosová
—
@zofkakosova

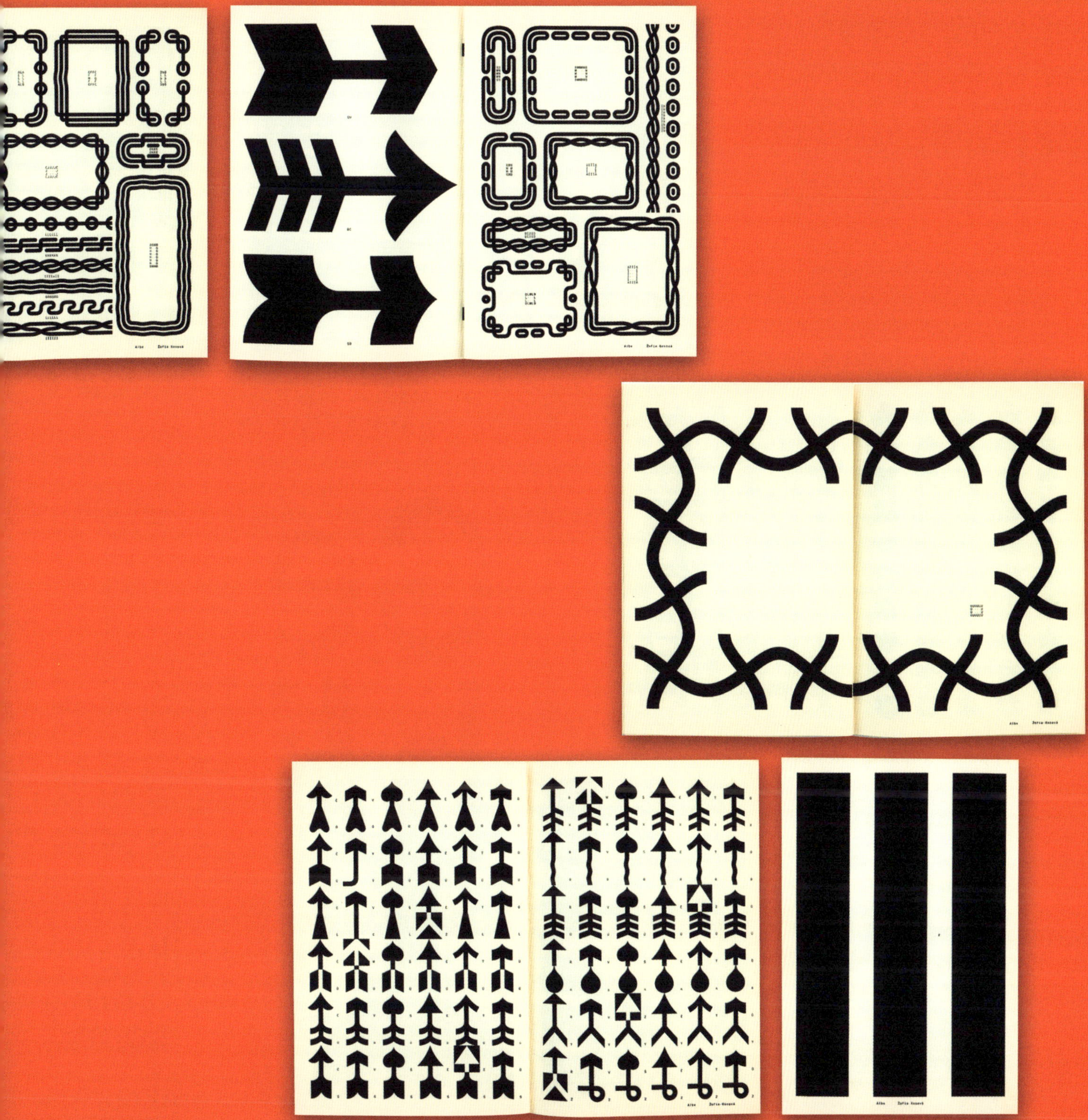

UMPRUM
Typeface: Several
Design: Several
—
umprum.cz @umprumtype

110 pt

Limit: Kintija Karala
Empe Sans: Regular, Italic
AbCcDdEeFfGgg
HhIiJjKLMmNnOo
PpQqRSsTtUVv
WwXxYyZz0123
456789.,:;!?/()
FREAKY 90s
ARE BACK
FOREVER
Gt'
Lia
Limit: Zofia Kosová
Chlebicheck: Regular, Medium
t
Chlebícheck
AaBDdEFfGgHI
JjKLMmNnOoPp
QqRrSsTtUuVv
WwXxYyZz012
3456789,.:!?&*
Element is one of the oldest Czech magazines, dealing with alternative culture and published almost completely independently of official publishers and grants. It began in MCMXCI as a response to cyberpunk magazines such as Mondo MM, graphic experimentation in Ray Gun magazine, and British dance magazines associated with the boom in the electronic scene in the XCs, such as i-D. Graphic designer and pedagogue Petr Babak talks about its origin as follows. The

— III —

3D Specimens

112 pt

This chapter is a fresh addition to the second edition of *Support Independent Type.* It explores type specimens presented on objects, merging typography with merchandising in inventive ways. From alphabet-clad ceramics to lettered apparel, these three-dimensional specimens are a playful and bold approach to showcasing fonts. It's typography you can touch, wear, or even drink from.

This phenomenon is part of a broader cultural trend where almost every community of enthusiasts creates its own universe of merchandise, turning objects into symbols of identity and belonging. These artifacts allow individuals to connect with their passions and signal their affiliations, transforming typography into a tangible sociological marker. 3D specimens are not just functional or decorative—they are storytelling tools and cultural badges. And let's face it, a scarf with a collection of letters from "A" to "Z" is simply warmer than a regular one!

NovoTypo
Typeface: Ziza Stencil
Design: Novo Typo / Mark van Wageningen
—
novotypo.nl @novotypo

114 pt

abcdefghij
klmnopqrs
tuvwxyz!?
0123456789

Twelve
Eleven One
Two
Three
Four
Five
Ten
Nine
Eight
Seven
Six
20 25 30 35 40
55 60 5

116 pt

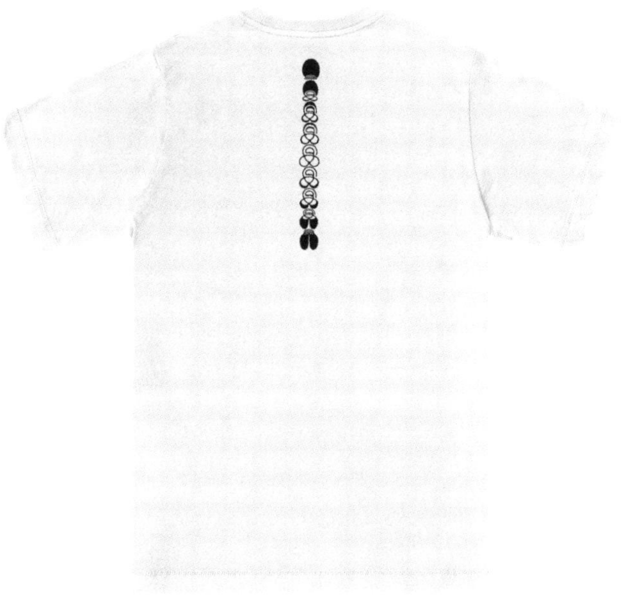

Universal Thirst
Typeface: Ilai
Design: Anagha Narayanan, Anurag Gautam
—
universalthirst.com @universalthirst

The Designers Foundry
Typeface: Tomato Grotesk, Romek
Design: Andrea Biggio
Clock designed by David Einwaller
—
thedesignersfoundry.com @tdfoundry

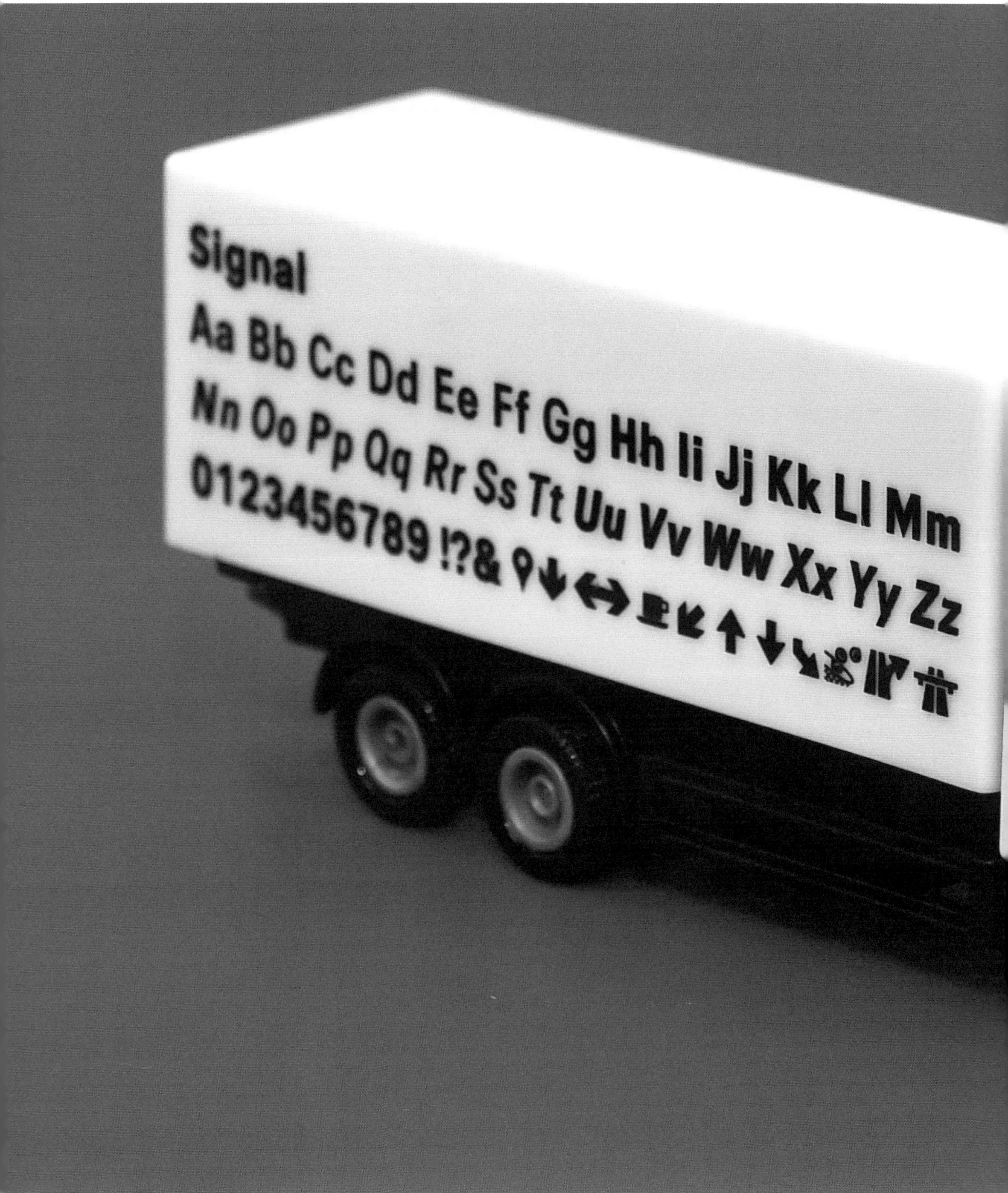
Signal
Aa Bb Cc Dd Ee Ff Gg Hh Ii Jj Kk Ll Mm
Nn Oo Pp Qq Rr Ss Tt Uu Vv Ww Xx Yy Zz
0123456789 !?&

118 pt

Production Type
Typeface: Siǵnal
Desiǵn: Emmanuel Besse
—
productiontype.com @productiontype

The Designers Foundry
Typeface: Halisa
Design: Jan Estrada-Osmycki
Keychain designed by Daniel McQueen
—
thedesignersfoundry.com @tdfoundry

AG Font
Typeface: Superblack Serif
Design: AG Font X Oncetype
—
agfont.com @ag.typography.institute

WiseType
Typeface: Saltburn
Design: Jacob Wise
—
wisetype.nl @wise_type

122 pt

Carefully designed for Track
Rail
WiseType® (2021)
2

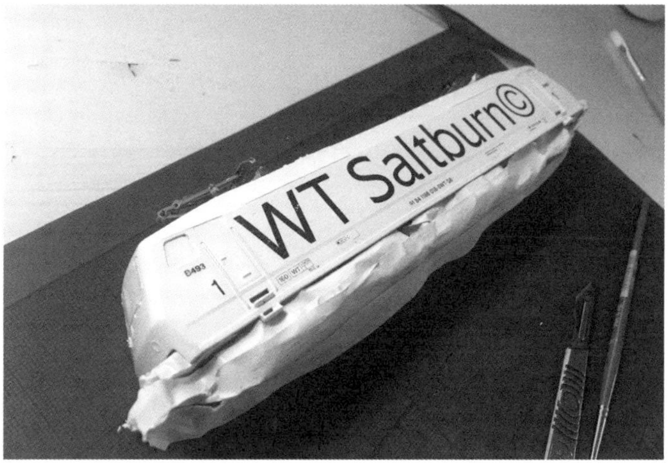
WT Saltburn©
1

110 pt
SPECIMEN
NOUVEAU
limoncello made in nancy alcohol – 35% • 25cl

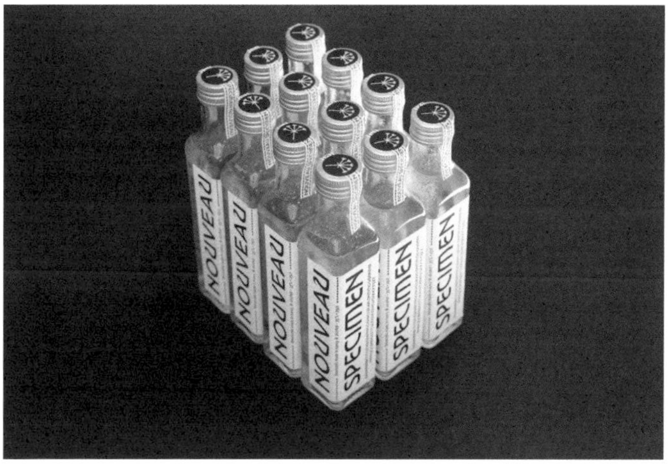

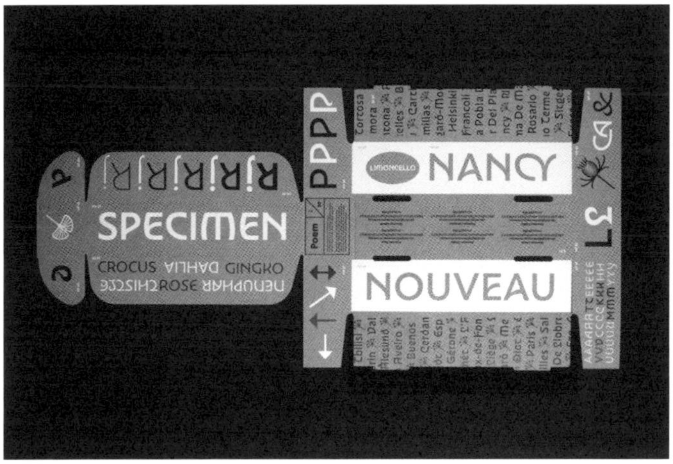

Poem Editions
Typeface: Nouveau
Design: Jérôme Knebusch,
Philippe Tytgat
—
poem-editions.com @poem_editions

P.S Type
Footnotes Socks
Design: P.S Type
—
pstypelab.com @markcaneso

Milletype Foundry
Typeface: MF Beach Club
Design: Milletype Foundry
—
milletype.fr @milletypefoundry

128 pt

Extraset
ES IBAN Scarf
Design: Extraset
—
extraset.ch @extraset.ch

Extraset
Typeface: ES Dokument
Design: Extraset
—
extraset.ch @extraset.ch

Peregrin Studio
Typeface: Declaro
Design: Peregrin Studio
—
peregrinstudio.com @peregrinstudio

130 pt

HOPE
LAD

Autograph
"Punktuation" T-shirt
Design: Peter Korsman
—
autograph.works @autograph__

Autograph
Drk Mttr 2.0
Design: Peter Korsman
—
autograph.works @autograph__

KOMETA Typefaces
KOMETA Scarf
Design: KOMETA
—
kometa.xyz @kometa.xyz

Order Type Foundry
Specimen Matchboxes
Design: Order Type Foundry
—
order.design/otf @ordertypefoundry

a
A B C D E F
G H I J K L
M N O P Q R
S T U V W X
Y Z a b c d
e f g h i j
k l m n o p
q r s t u v
w x y z 0 1
2 3 4 5 6 7
8 9 ! ? @ $
.otf
Order Type Foundry
order.design/otf

A B C D E F
G H I J K L
M N O P Q R
S T U V W X
Y Z a b c d
e f g h i j
k l m n o p
q r s t u v
w x y z 0 1
2 3 4 5 6 7
8 9 ! ? @ $
.otf

b

C
A B C D E F
G H I J K L
M N O P Q R
S T U V W X
Y Z a b c d
e f g h i j
k l m n o p
q r s t u v
w x y z 0 1
2 3 4 5 6 7
8 9 ! ? @ $
.otf
Order Type Foundry
order.design/otf

WELTKERN® Typefaces
Typeface: Burns
Design: Nolan Paparelli
—
weltkern.com @type.weltkern

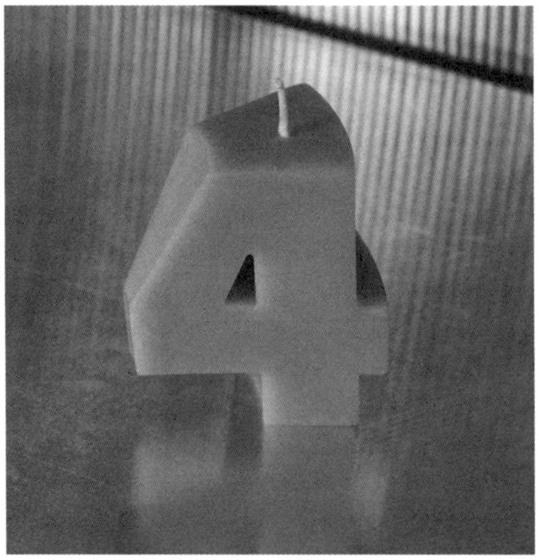

Milletype Foundry
Typeface: MF Corinthe
Design: Romain Marc
—
milletype.fr @milletypefoundry

Yota Fonts
Typeface: Tuileries
Design: Yorgo Tloupas, Martin Pasquier
—
yotafonts.com @yotafonts

Yota Fonts
Typefaces: Hypercube (Vase), Lipslide (Wax Candle), Tuileries (Candle Holder), Nocturne (Ashtray)
Design: Yota Fonts
—
yotafonts.com @yotafonts

TUILERIES

Yota Fonts
Typeface: Tuileries
Design: Yorgo Tloupas, Martin Pasquier
—
yotafonts.com @yotafonts

Nocturne
Medium
Bold
Regular
Bold Italic
Medium Italic
abCde
12345
67890

Yota Fonts
Typeface: Nocturne
Design: Yorgo Tloupas, Martin Pasquier
—
yotafonts.com @yotafonts

Yota Fonts
Typeface: Hypercube
Design: Yorgo Tloupas, Martin Pasquier
—
yotafonts.com @yotafonts

Threedotstype
Typeface: Di Grotesk
Design: Marian Misiak
—
threedotstype.com @threedotstype

146 pt

12 pt
9 pt
6 pt
3 pt
1 pt

148 pt

Support Independent Type II

AllCaps
Typeface: Modern Gothic Regular
Design: Malte Bentzen
Cap designed by AllCaps
—
allcapstype.com @allcapstype

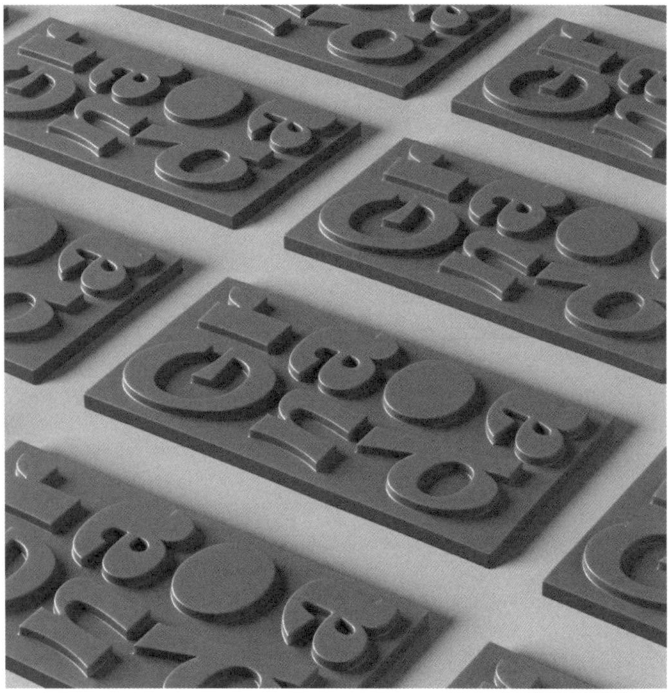

The Designers Foundry
Granda Chocolate Bar
Typeface: Granda
Design: Radek Łukasiewicz
—
thedesignersfoundry.com @tdfoundry

The Designers Foundry
Goodman Blanket
Typeface: Goodman
Design: Ethan Nakache
—
thedesignersfoundry.com @tdfoundry

— IV —
Instagram

152 pt

Love it or hate it, social media shapes the way we experience design today. It dictates trends, compresses complex ideas into shareable moments, and often prioritizes visibility over depth. While it can foster connection and amplify voices, it could also risks reducing creative work to fleeting impressions and algorithm-driven popularity contests. This chapter dives into how independent type creators use platforms like Instagram to promote their work. We've organized these examples into categories to showcase the variety, ingenuity, and sheer creativity of their approaches. It's proof that even a grid of tiny squares can be both a powerful stage for typographic storytelling and a challenging space for meaningful engagement.

Margot Lévêque Type Foundry
Typeface: Ninna
Design: Margot Lévêque
—
@margot.leveque

Formula Type
Typeface: FT Habit
Design: Piero Di Biase
—
@formula.type

Fatype
Typeface: Laplace Mono
Design: Anton Koovit
—
@fatypefonts

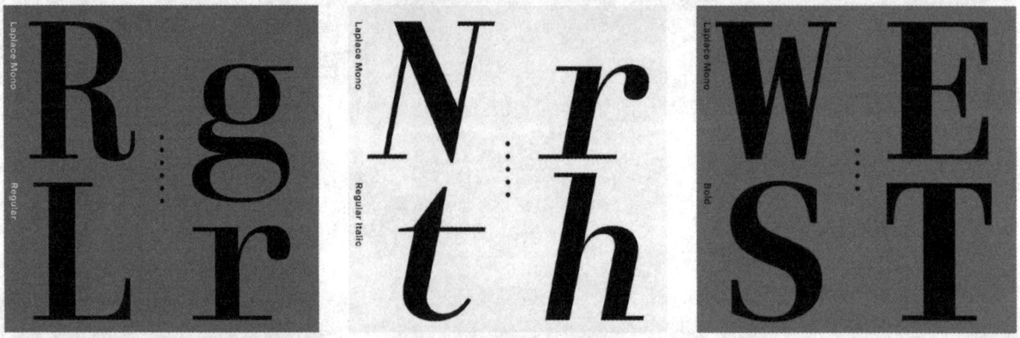

Brownfox
Typeface: Formular Neutral
Design: Vyacheslav Kirilenko, Gayaneh Bağdasaryan
—
@brownfoxtype

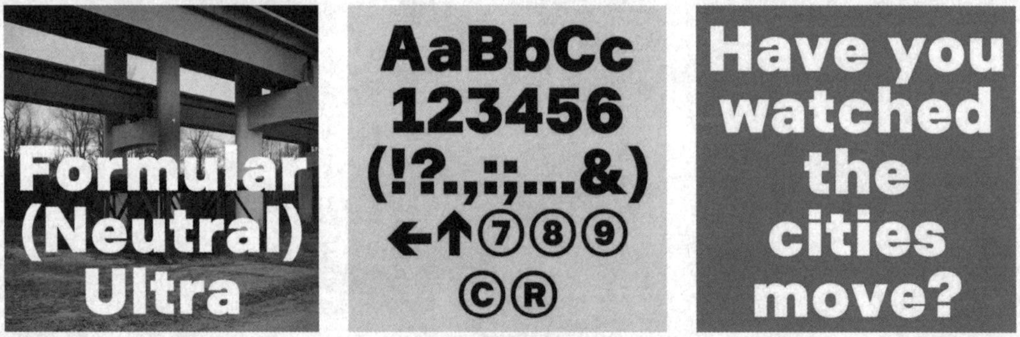

Mass-Driver
Typeface: Polychrome
Design: Mass-Driver
—
mass-driver.com

154 pt

(1) Carousel

STEROID
BROWNIE
FRUITY
CRUMMY
YUMMY

Vous prendrez plaisir
à n'importe quelle activité
à laquelle vous serez
totalement présent,
à n'importe quelle activité
ne représentant pas
seulement un moyen
pour arriver à
une fin.

"Smoke" was released in 1995 and marked a collaboration between director Wayne Wang and writer Paul Auster. Auster adapted the screenplay from his own short stories, which were originally published in his collection titled *"The New York Trilogy."* The film's narrative weaves together multiple vignettes, each focusing on different characters whose lives intersect at the Brooklyn tobacco shop owned by Auggie Wren, portrayed by Harvey Keitel.

B'klyn, Nyc

Dű* ** ďčħ

Elvis Presley
Banana and Peanut
Butter Sandwich
*1935 – 1977 †

Marilyn Monroe
Hot Baths
*1926 – 1962 †

Muhammad Ali
Rhymes and Poems
*1942 – 2016 †

Audrey Hepburn
Gardening
*1929 – 1993 †

Laplace Mono

Bold Italic

B a b e

Laplace Mono

Light

P a c k

I with thee, and thou with me

(1) Carousel

Boğidar Mascareñas
Typeface: Ager
Design: Boğidar Mascareñas
—
@boğidarmascarenas

Frank Fonts
Typeface: Kraken
Design: Béla Frank
—
@frank_fonts

Out of the Dark
Typeface: Cosplay
Design: Out of the Dark
—
@outofthedark.xyz

Blast Foundry
Typeface: Bizzarrini
Design: Diana Ovezea
—
@blast_foundry

Bruta Types
Typeface: Monor
Design: Bruta Types
—
@bruta.types

156 pt

(1) Carousel

(1) Carousel

Frank Fonts
Typeface: Semafor
Design: Frank Fonts
—
@frank_fonts

PFA Typefaces
Typeface: Knochen
Design: Martin Aleith
—
@pfa_typefaces

Edition Studio
Typeface: Syndicat
Design: Adrien Menard
—
@edition.studio

bb-bureau
Typeface: Waldeck
Design: Benoît Bodhuin
—
@benoitbodhuin

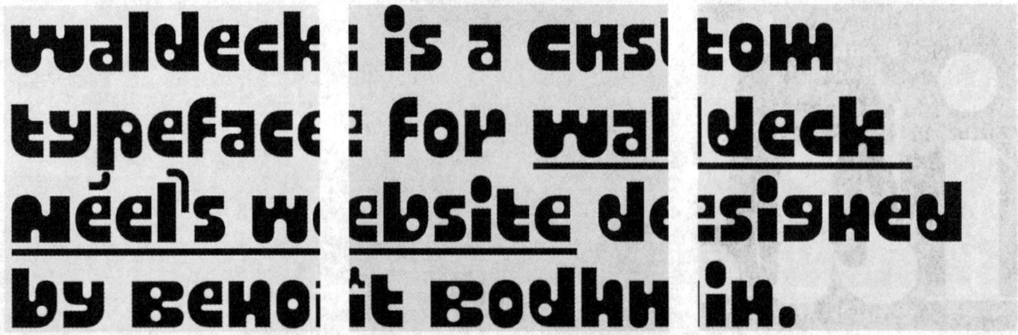

WiseType
Typeface: Zaft²
Design: Jacob Wise, Céline Hurka
—
@wise_type

158 pt

(1) Carousel

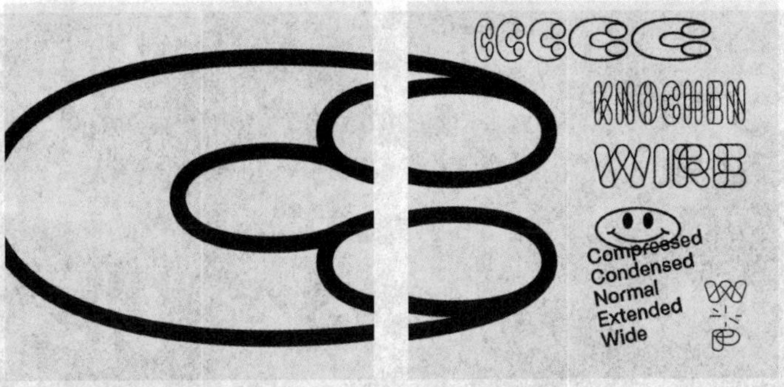
KNOCHEN
WIRE
Compressed
Condensed
Normal
Extended
Wide

HOT
DOG
Light

(1) Carousel

Universal Thirst
Typeface: Ilai
Design: Universal Thirst
—
@universalthirst

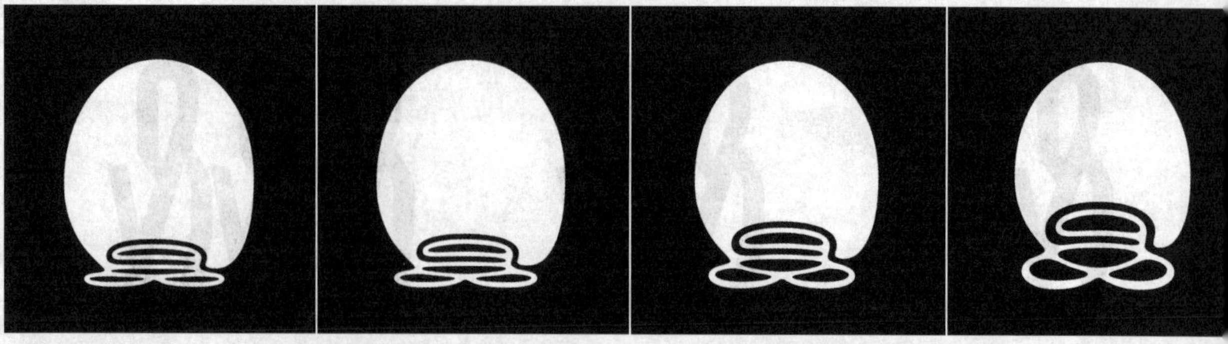

Lift Type
Typeface: Gustavo
Design: Romain Oudin
—
@lifttype

Monkey Type
Typeface: Jane
Design: Monkey Type
—
@mnkytype

bvh type
Typeface: Eddi
Design: bvh type
—
@bvhtype

Tightype
Typeface: Macan
Design: Tightype
—
@tightype

(2) Motion

Optimo
Typeface: Basel
Design: Chi-Long Trieu
—
@optimofoundry

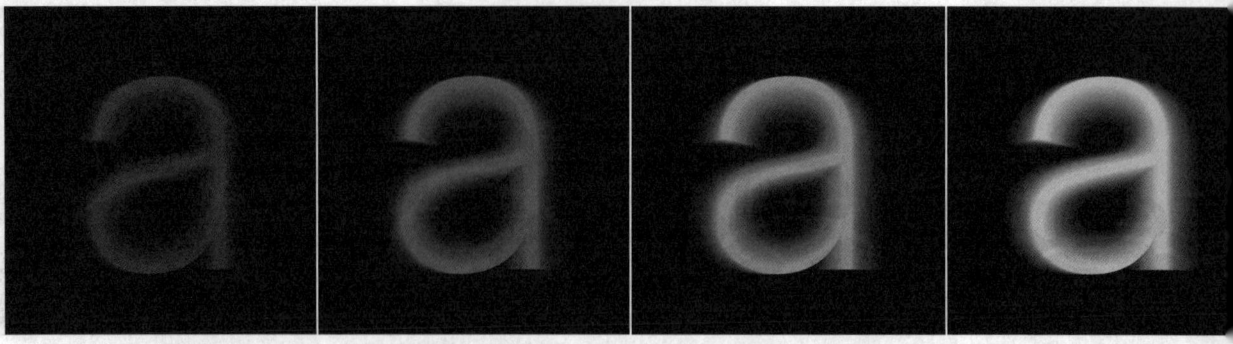

NaN
Typeface: Jaune
Design: Jérémy Landes
—
@nan_xyz

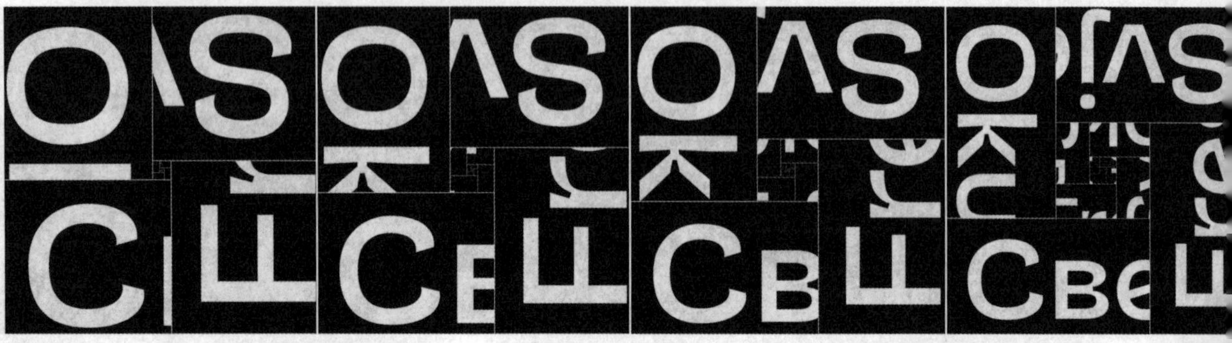

Dinamo Typefaces
Typeface: Social
Design: Dinamo
—
@abcdinamo

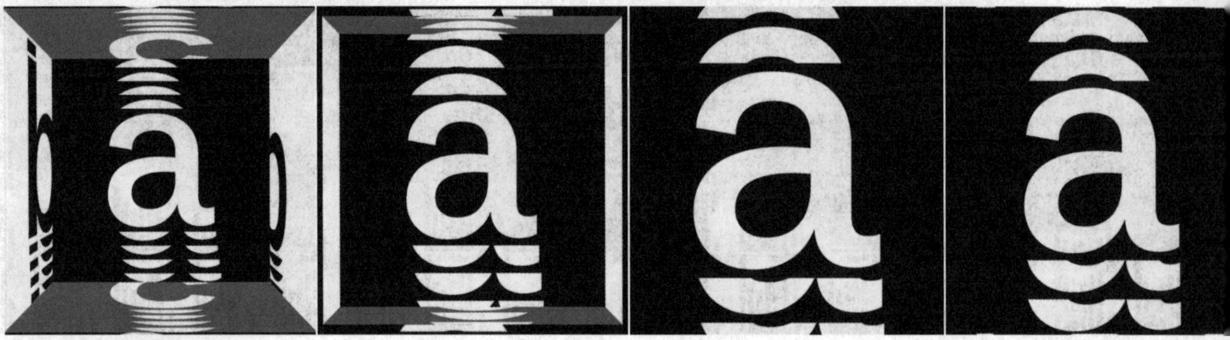

Central Type
Typeface: Dirk
Design: Mark Butchko
—
@central.type

Monkey Type
Typeface: Louie
Design: Monkey Type
—
@mnkytype

(2) Motion

— IV —

Instagram Single

164 pt

Metis Foundry
Typeface: Odd Mono
Design: Simon Bent
—
@metis_foundry

Swiss Typefaces
Typeface: TheW Clan ODB
Design: Ian Party / Swiss Typefaces
—
@swisstypefaces

Briefcase Type Foundry
Typeface: Ludva
Design: Marek Čuban
—
@briefcasetype

166 pt

(1) Pictorial Type

Nouvelle Noire
Typeface: Pivot Grotesk
Design: Anton Studer
—
@nouvellenoire_ch

Parachute
Typeface: Regal Stencil
Design: Panos Vassiliou
—
@parachutetypefoundry

[1] **Quicknap.zzz**
Typeface: Gosna
Design: Dae Huen Lee
@quicknap.zzz

[2] **Hey–Type**
Typeface: 800G
Design: Jazlyn, Fung Hei Man
@jazheiman

[3] **Type of Feeling**
Typeface: Satori
Design: Type of Feeling
@typeoffeeling

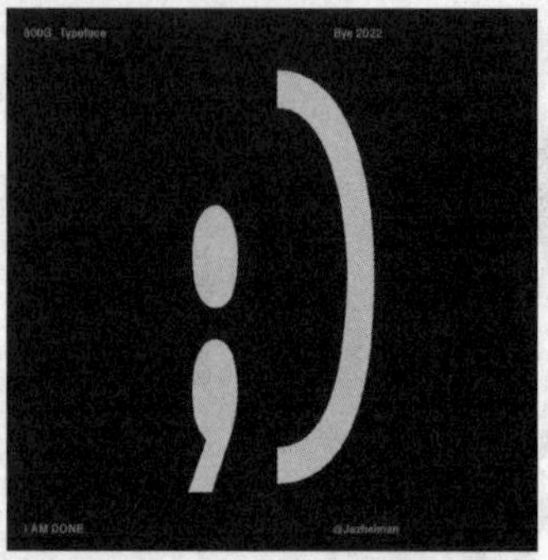

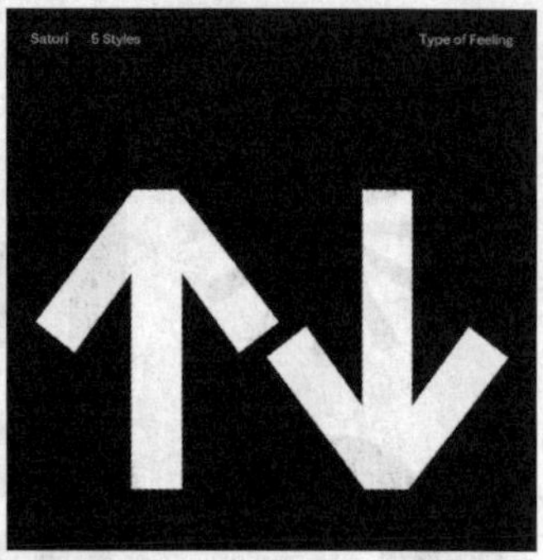

[1] **Poem Editions**
Typeface: Nouveau
Design: Jérôme Knebusch,
Philippe Tytgat
@poem_editions

[2] **Brass Fonts**
Typeface: Garant
Design: Guido Schneider
@brassfonts

[3] **Newglyph**
Typeface: NeoGeo
Design: Ian Party
@newglyph

[1] **WiseType**
Typeface: Gablet
Design: Jacob Wise
@wise_type

[2] **Giulia Boggio**
Typeface: Bugia
Design: Giulia Boggio
@bojjoe

[3] **Bold Studio**
Typeface: Casual
Design: Bold Studio
@bold.studio

[1] **Fuerte Type**
Typeface: Sakral
Design: Fermin Guerrero
@fuertetype

[2] **Naipe Foundry**
Typeface: Discordia
Design: Naipe Foundry
@naipe.xyz

[3] **Monomodular Type Foundry**
Typeface: Gradial
Design: Lena Weber
@lenaweber404

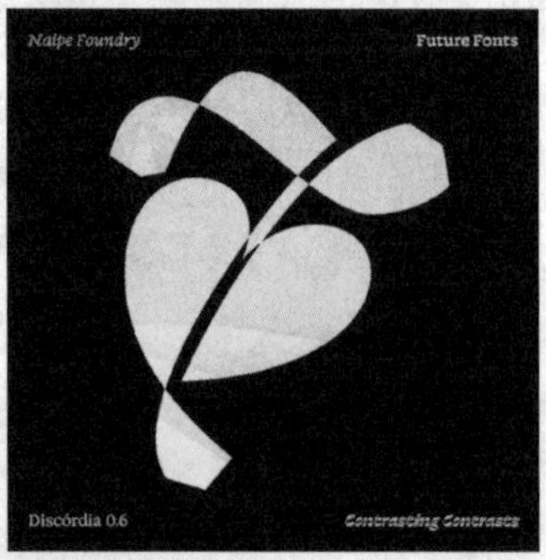

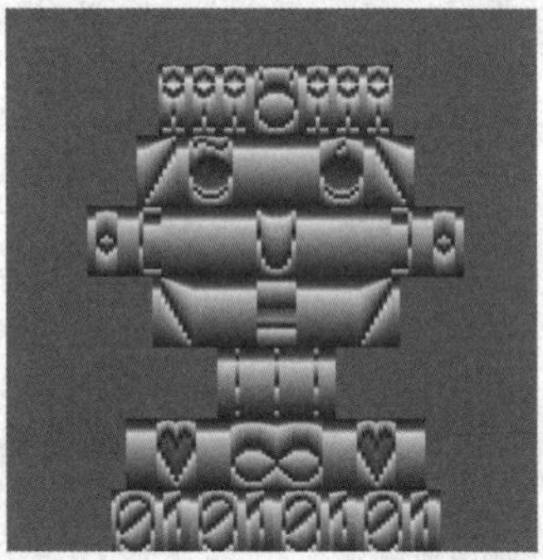

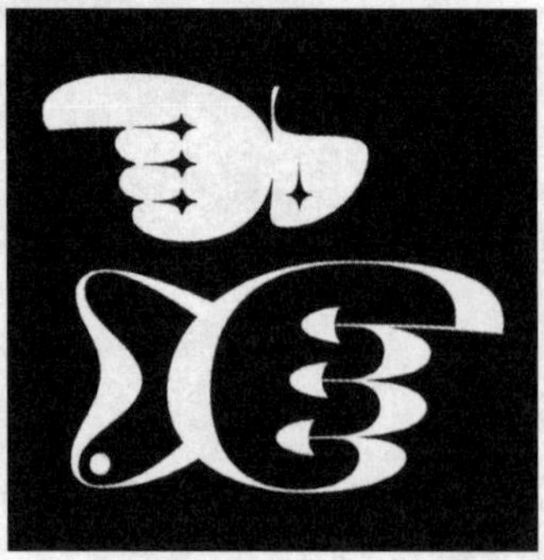

[2] **Character Type**
Typeface: News Serif
Design: Henning Skibbe
@character.type

[3] **Tightype**
Typeface: Habitas
Design: Fabian Fohrer,
Léa Bruneau
@tightype

168 pt

(1) Pictorial Type

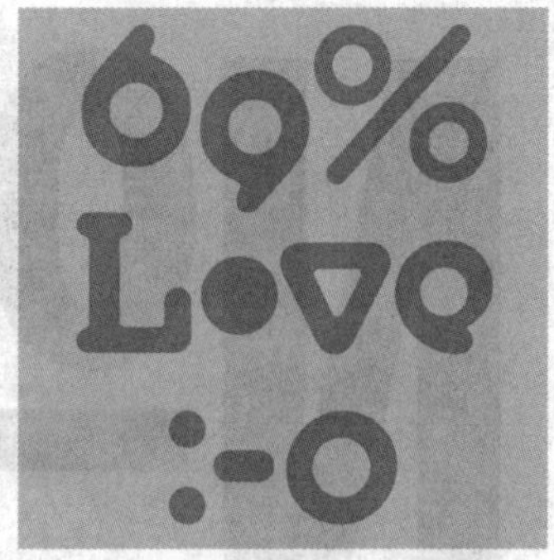

[1] **Diorama Type Partners**
Typeface: Mesh
Design: Marie-Mam Sai Bellier
@dioramatypepartners

[2] **Altiplano**
Typeface: Nirvana
Design: Altiplano
@altiplano_typefaces

[3] **Kanon Foundry**
Typeface: Clippy
Design: Tor Weibull
@kanon_foundry

[1] **Lift Type**
Typeface: Baste
Design: Image Format
@lifttype

[2] **PLAYTYPE**
Typeface: South East
Design: Jonas Hecksher
@playtype

[3] **Fantasia Type**
Typeface: Cipria
Design: Fantasia Type
@fantasia.type

[1] **Brownfox**
Typeface: Ferro
Design: Vyacheslav Kirilenko, Gayaneh Bagdasaryan
@brownfoxtype

[2] **Apex Type Foundry**
Typeface: Practical Grotesk
Design: Alex Chavot
@apex_type_foundry

[3] **HAGILDA**
Typeface: Skater
Design: HAGILDA
@hagilda

[1] **Boafff**
Typeface: Drago
Design: Boafff
@boafff

[2] **Extraset**
Typeface: Rebond Grotesque
Design: Roger Gaillard
@extraset.ch

[3] **Tunera Type Foundry**
Typeface: Picaflor
Design: Ariel Martín Pérez
@tunera.xyz

[1] **atipo foundry**
Typeface: Novela
Design: atipo foundry
@atipostudio

[2] **bvh type**
Typeface: BLine
Design: André Baldinger, Fanny Hamelin
@bvhtype

(1) Pictorial Type

ALT.tf
Typeface: ALT Riviera
Design: Giulia Boggio
—
@alt_tf

Monkey Type
Typeface: Banana Mono
Design: Monkey Type
—
@mnkytype

In-House Intl
Typeface: Broker
Design: In-House Intl
—
@weareinhouse

170 pt

(2) Loud Type

Jung-Lee Type Foundry
Typeface: Impact Nieuw
Design: Jungmyung Lee
—
@jung_lee_type_foundry

Apex Type Foundry
Typeface: Nicéphore
Design: Apex Type Foundry
—
@apex_type_foundry

Feliciano Type
Typeface: Parafina
Design: Feliciano Type
—
@felicianotype

(2) Loud Type

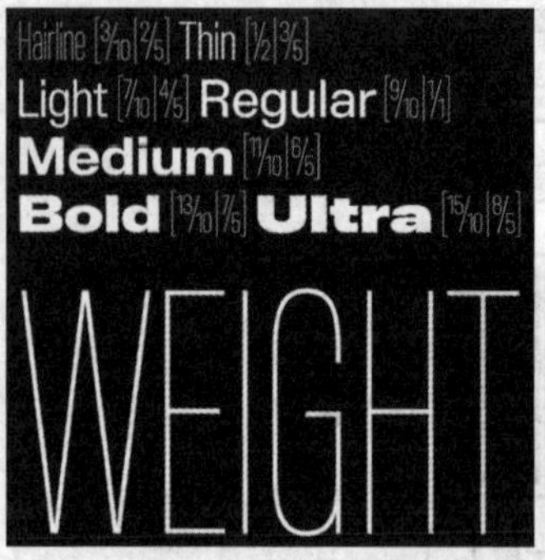

[1] **Charlotte Rohde**
Typeface: New Edge 666
Design: Charlotte Rohde
@charlotte__rohde

[2] **Bold Studio**
Typeface: BB Torsos
Design: Bold Studio
@bold.studio

[3] **Optimo**
Typeface: Apax
Design: François Rappo
@optimofoundry

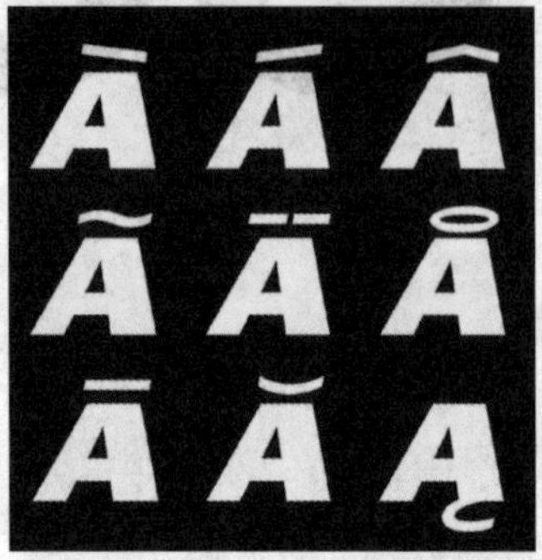

[1] **Briefcase Type Foundry**
Typeface: Orion
Design: Stanislav Maršo
@briefcasetype

[2] **Frank Fonts**
Typeface: Bravour
Design: Frank Fonts
@frank_fonts

[3] **Power Type**
Typeface: Power Display
Design: Teğuh Arief
@powertypefoundry

[1] **Central Type**
Typeface: Zonk
Design: Mark Butchko
@central.type

[2] **CoType Foundry**
Typeface: Scandium
Design: CoType Foundry
@cotypefoundry

[3] **Character Type**
Typeface: Tragic Grotesk
Design: Henning Skibbe
@character.type

[1] **Erkin Karamemet**
Typeface: Modena
Design: Erkin Karamemet
@erkin_karamemet

[2] **Alaa Tameem**
Typeface: The Manchette
Design: Alaa Tameem
@alaatameem

[3] **Contrast Foundry**
Typeface: CoFo Gothic
Design: Maria Doreuli
@contrastfoundry

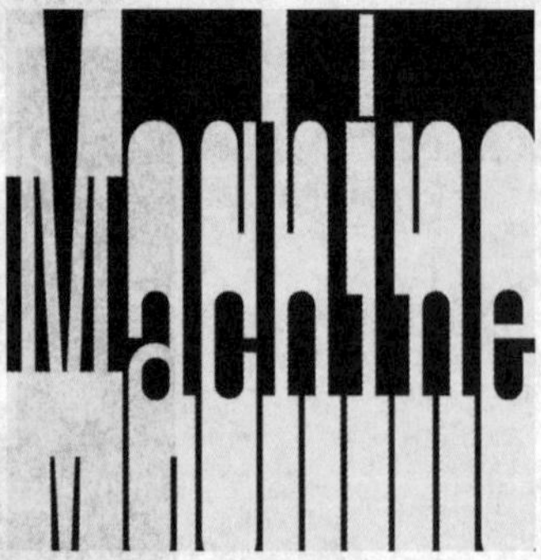

[2] **David Jonathan Ross**
Typeface: Pappardelle
Design: David Jonathan Ross
@djrrb

[3] **RadLuka**
Typeface: RL Limo
Design: Radek Łukasiewicz
@radek_lukasiewicz

[1] **APFEL Type Foundry**
Typeface: Certeau
Design: APFEL Type Foundry
@apfeltypefoundry

[2] **AllCaps**
Typeface: Bandit
Design: Sylvan Lanz
@allcapstype

[3] **Love Letters Typefaces**
Typeface: Untitled
Design: Love Letters Typefaces
@sebastien_sanfilippo

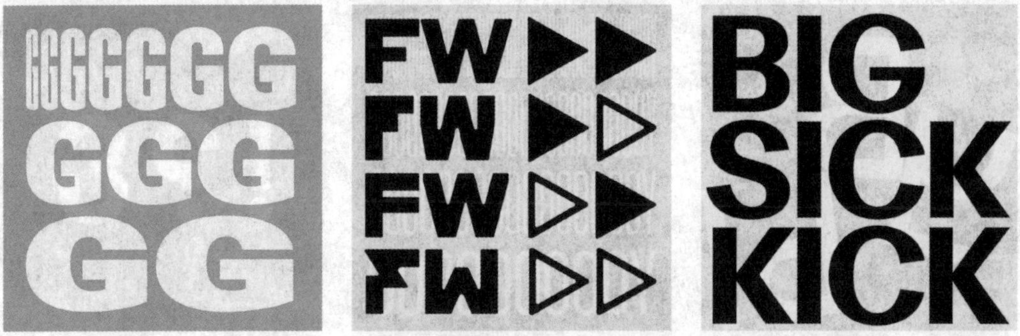

[1] **Dinamo Typefaces**
Typeface: Gravity
Design: Dinamo
@abcdinamo

[2] **HAGILDA**
Typeface: Levit
Design: HAGILDA
@hagilda

[3] **Kanon Foundry**
Typeface: Diagol Grotesk
Design: Tor Weibull
@kanon_foundry

[1] **The Letters**
Typeface: Fiction
Design: Mark Gowing
@theletters.co

[2] **In-House Intl**
Typeface: Vartek
Design: Inhouse Type
@inhousetype

[3] **bvh type**
Typeface: BVH BLine
Design: André Baldinger, Fanny Hamelin
@bvhtype

[1] **DSType Foundry**
Typeface: Pratico Slab
Design: DSType Foundry
@dstype_foundry

[2] **Heavyweight**
Typeface: Animo
Design: Heavyweight
@heavyweight_type

[3] **Or Type**
Typeface: Forzata
Design: Mads Freund Brunse
@ortype

[1] **Peregrin Studio**
Typeface: Systemia
Design: Peregrin Studio
@peregrinstudio

[2] **Juri Zaech**
Typeface: OBO Star
Design: Juri Zaech
@jurizaech

[1] **Rüdiger**
Typeface: Yuni Slab
Design: Philipp Neumeyer
@rudiger.xyz

[2] **Newglyph**
Typeface: Aloha
Design: Ian Party
@newglyph

[3] **Boğidar Mascareñas**
Typeface: Ager
Design: Boğidar Mascareñas
@bogidarmascarenas

[1] **Shriftovik**
Typeface: SK Concretica
Design: Andy Karter, Tikhon Reztcov
@shriftovik

[2] **Arillatype**
Typeface: AT Slam
Design: Pedro Arilla
@arillatype

[3] **So Type**
Typeface: Hyra Sans
Design: So Type
@sotypefoundry

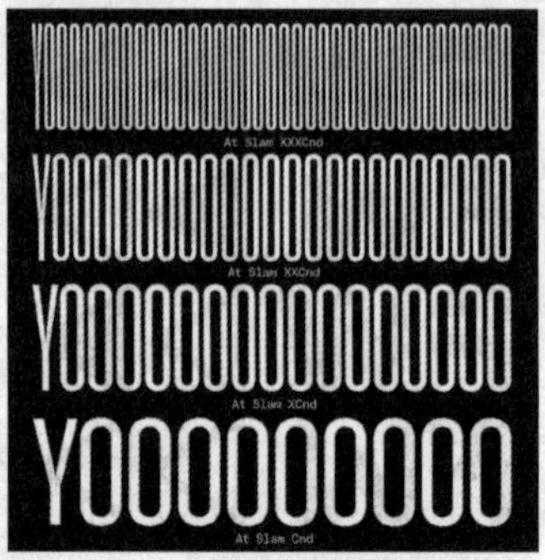

[1] **Vectro Type Foundry**
Typeface: Rainer
Design: Philipp Neumeyer
@vectrotype

[2] **Displaay**
Typeface: Denim
Design: Displaay
@xyz_displaay

[3] **Nuform Type**
Typeface: Ozik
Design: Erik Marinovich
@nuformtype

[1] **Underware**
Typeface: Plakto
Design: Underware
@underwarefoundry

[2] **WELTKERN® Typefaces**
Typeface: Everett
Design: Nolan Paparelli
@type.weltkern

[3] **Polytype**
Typeface: Gela
Design: Lewis MacDonald
@polytypefoundry

[2] **Hypertype**
Typeface: Neutronic
Design: Hypertype
@hypertype.xyz

[3] **Universal Thirst**
Typeface: Ouma
Design: Gunnar Vilhjálmsson, Hitesh Malaviya, Gabriel Markan
@universalthirst

(2) Loud Type

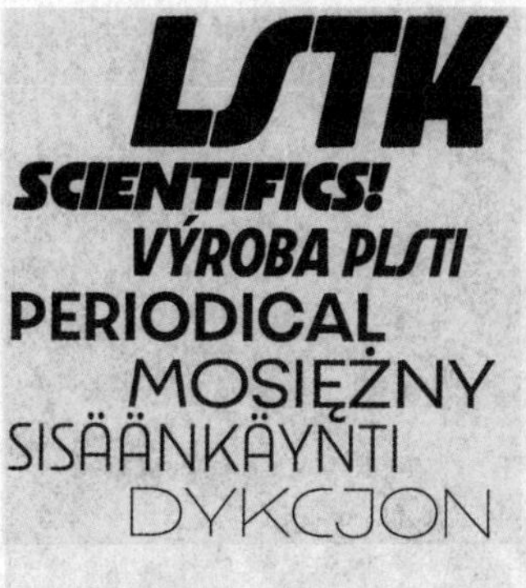

[1] **Ty Finck**
Typeface: Furrow
Design: Ty Finck
@tyfromtheinternet

[2] **Process Type Foundry**
Typeface: Landform
Design: Eric Olson
@process_type

[3] **East of Rome**
Typeface: Komplekt
Design: Lewis McGuffie
@eastofrome

[1] **Arillatype**
Typeface: AT Slam
Design: Pedro Arilla
@arillatype

[2] **Estudio Calderón**
Typeface: Ocelot
Design: Felipe C. Arteaga
@calderon_type

[3] **Modern Type**
Typeface: Mighty Mono
Design: Thy Hà
@moderntype_

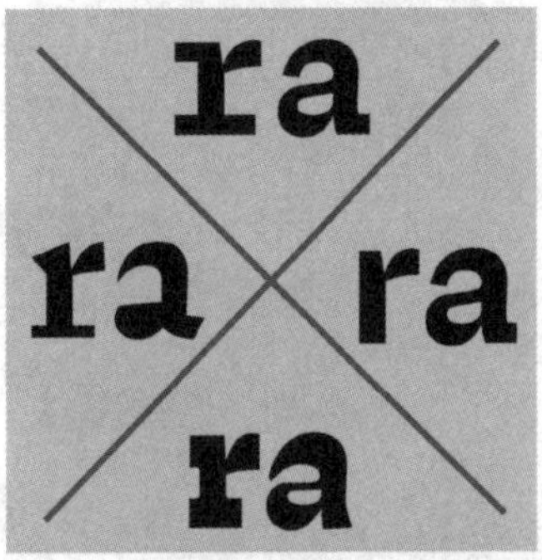

[1] **Phantom Foundry**
Typeface: Token
Design: Jean-Baptiste Morizot
@phantomfoundry

[2] **Erkin Karamemet**
Typeface: Modena
Design: Erkin Karamemet
@erkin_karamemet

[3] **W Type Foundry**
Typeface: Puffy
Design: Gaspar Muñoz
@wtypefoundry

[1] **Typeverything**
Typeface: Bourbon St
Design: Andrei Robu
@typeverything

[2] **Sumotype**
Typeface: Gregor
Design: Oscar Guerrero Cañizares
@sumotype

[3] **APFEL Type Foundry**
Typeface: Remnants
Design: APFEL Type Foundry
@apfeltypefoundry

[1] **26A1**
Typeface: Marund
Design: Daniel Wenzel
@studio26a1

[2] **Zin Nagao**
Typeface: ZNVT24
Design: Zin Nagao
@zinnagao

WELTKERN® Typefaces
Typeface: Ghost
Design: Nolan Paparelli
—
@type.weltkern

Arillatype
Typeface: AT Aero
Design: Pedro Arilla
—
@arillatype

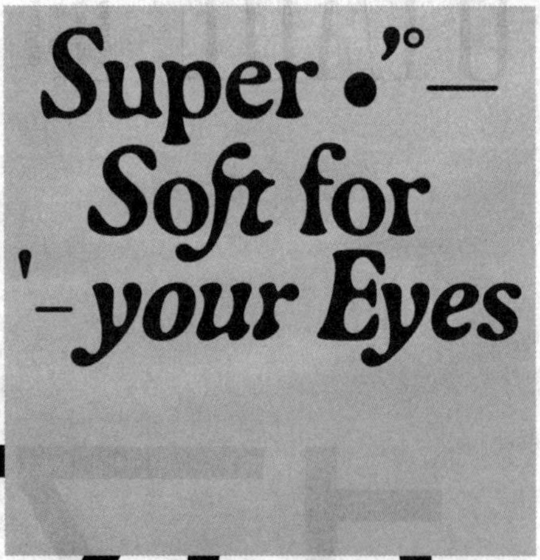

Out of the Dark
Typeface: Gaya
Design: Raphaël de La Morinerie
—
@outofthedark.xyz

176 pt

(3) Calm Type

Margot Lévêque Type Foundry
Typeface: Ninna
Design: Margot Lévêque
—
@margot.leveque

Bureau Brut
Typeface: Round
Design: Daniela Party
—
@bureau_brut

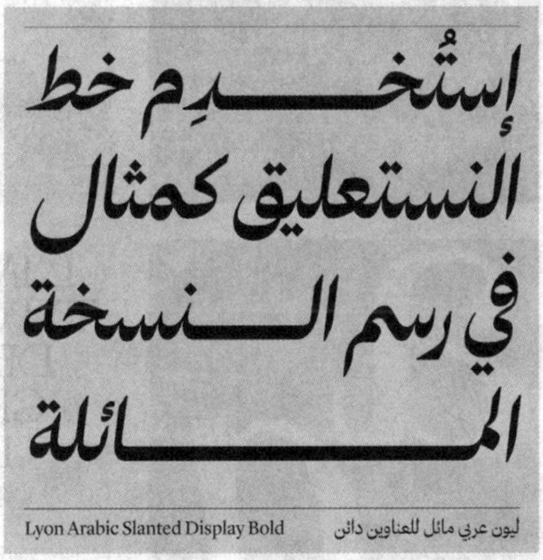

Commercial Type
Typeface: Lyon
Design: Kai Bernau, Ilya Ruderman, Irene Vlachou
—
@commercialtype

[1] **Slow Fonts**
Typeface: Method Mono
Design: Fabien Coupas
@slowfonts

[2] **Tunera Type Foundry**
Typeface: Nyght Serif
Design: Maksym Kobuzan
@tunera.xyz

[3] **Flavia Zimbardi**
Typeface: Hanae
Design: Flavia Zimbardi, Ayaka B.
@flaviazim

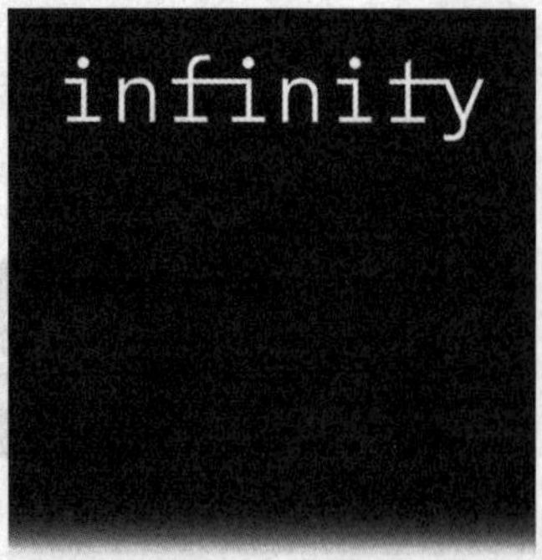

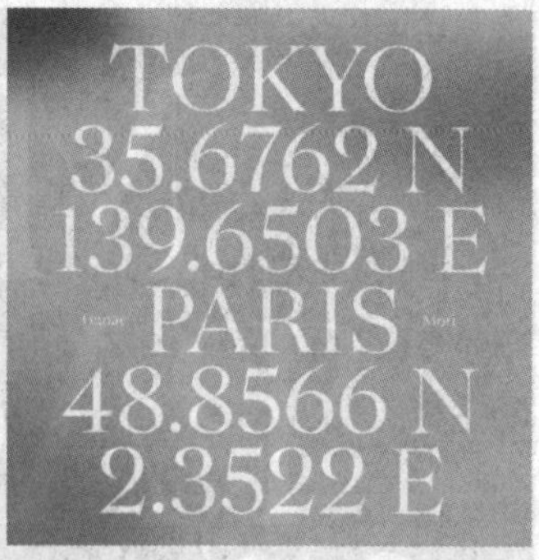

[1] **Hurme Design**
Typeface: Hurme Oval Sans
Design: Hurme Design
@hurmedesign

[2] **So Type**
Typeface: KPA Serif Display
Design: So Type
@sotypefoundry

[3] **Naipe Foundry**
Typeface: Discordia
Design: Naipe Foundry
@naipe.xyz

[1] **Sharp Type**
Typeface: Sharp Roman
Design: Sharp Type
@sharp_type

[2] **The Northern Block**
Typeface: Angusta
Design: The Northern Block
@northernblock

[3] **Daytona Mess**
Typeface: Nautila
Design: Anne-Dauphine Borione
@daytonamess.otf

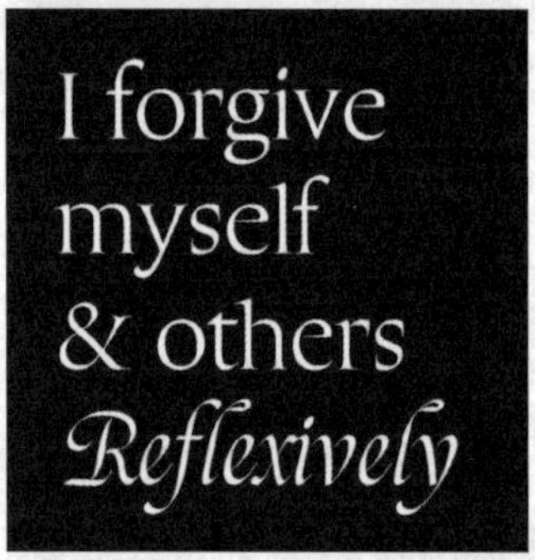

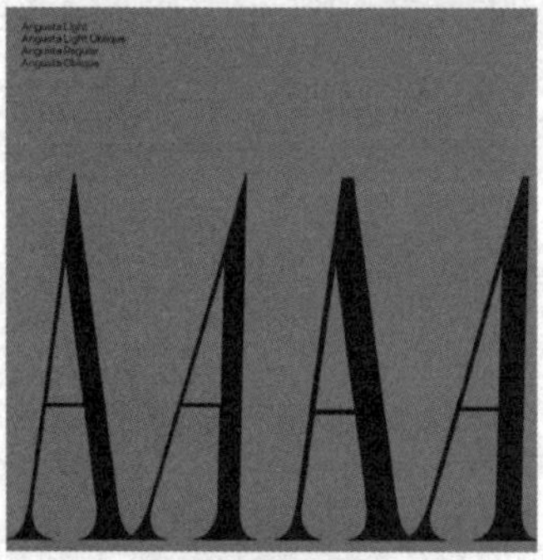

[1] **205TF**
Typeface: Exposure
Design: Federico Parra Barrios
@205tf

[2] **Beasts of England**
Typeface: Newsagent
Design: Beasts of England
@beastsofengland

[3] **Ultra Kuhl**
Typeface: Alyc
Design: Ultra Kuhl
@ultrakuhl

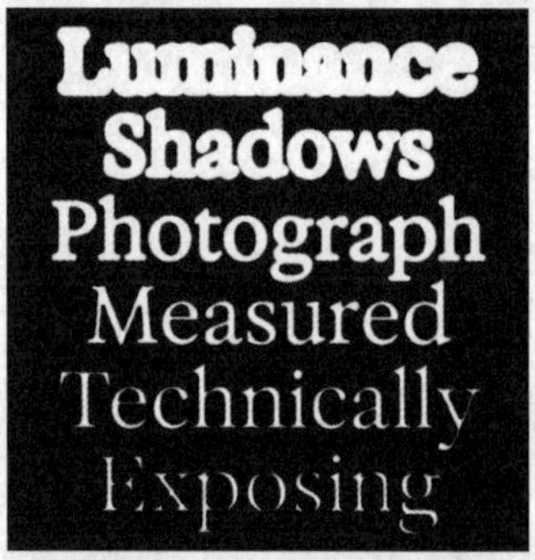

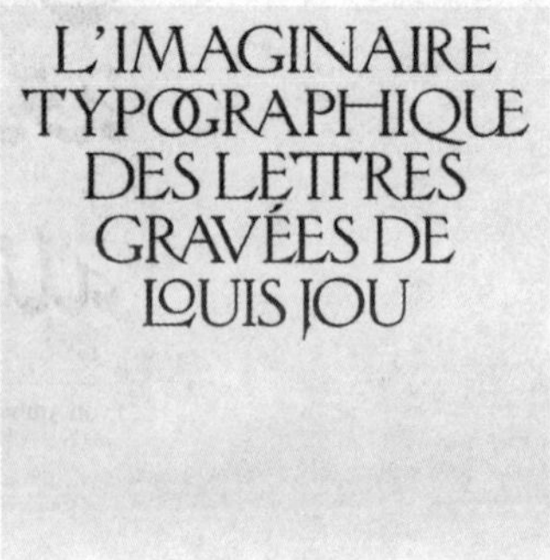

[2] **ECAL Typefaces**
Typeface: Mule
Design: Samira Schneuwly
@ecalmatd

[3] **Jules Durand**
Typeface: Jou
Design: Jules Durand
@lazy_dog.ttf

170 pt

(3) Calm Type

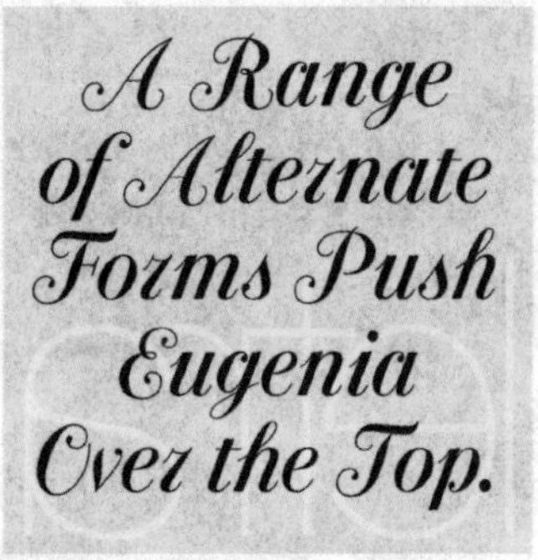

[1] **Bureau Brut**
Typeface: Boogy Brut
Design: Bureau Brut
@bureau_brut

[2] **Type Salon**
Typeface: Wesna
Design: Type Salon
@typesalon

[3] **Commercial Type**
Typeface: Eugenia
Design: Miguel Reyes
@commercialtype

[1] **Schick Toikka**
Typeface: Apris
Design: Schick Toikka
@schicktoikka

[2] **Edition Studio**
Typeface: Syndicat Grotesk
Design: Edition Studio
@edition.studio

[3] **Or Type**
Typeface: Lemmen
Design: Or Type
@ortype

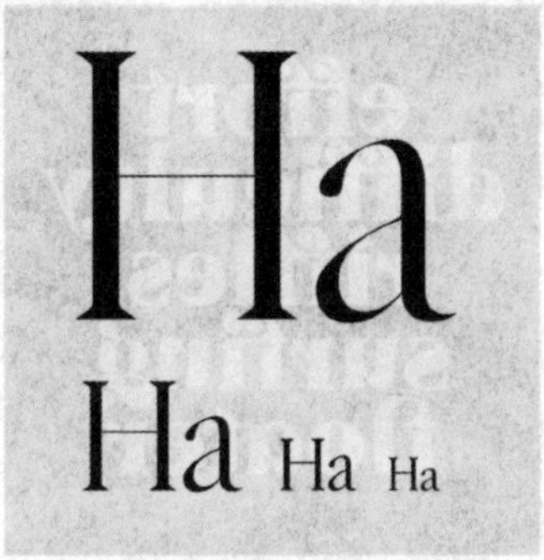

[1] **Bouk Ra**
Typeface: Hanol
Design: Bouk Ra
@_bouk_ra_

[2] **Feliciano Type**
Typeface: Parnaso
Design: Feliciano Type
@felicianotype

[3] **Typografische**
Typeface: Alpha
Design: Typografische
@typografische

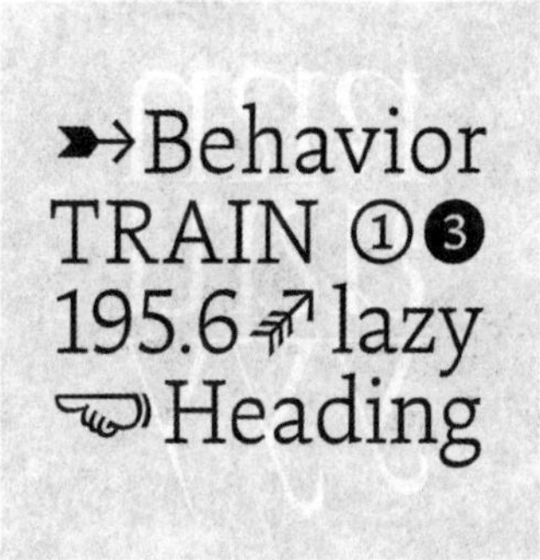

[1] **Family Type**
Typeface: Aime
Design: Family Type
@familytype.co

[2] **Polytype**
Typeface: Freizeit
Design: Polytype
@polytypefoundry

[3] **Jan Fromm**
Typeface: Capito
Design: Jan Fromm
@jan__fromm

[1] **Eko Bimantara**
Typeface: Realitas
Design: Eko Bimantara
@ekobimantara_fonts

[2] **Boulevard LAB**
Typeface: Sometimes Neue
Design: Boulevard LAB
@boulevardtype

(3) Calm Type

[1] **Maxitype**
Typeface: Selecta
Design: Daniel Haettenschwiller
@maxitype_com

[2] **Nodo Type Foundry**
Typeface: NT Neuss
Design: Nodo Type Foundry
@nodotypefoundry

[3] **Superior Type**
Typeface: Civil & Epika Serif
Design: Civil – Matyáš Machat,
Epika Serif – Rostislav Vaněk
@superior_type

[1] **Duong Tran**
Typeface: Loes
Design: Duong Tran
@dgtran.otf

[2] **The Northern Block**
Typeface: Loew Next
Design: The Northern Block
@northernblock

[3] **Heavyweight**
Typeface: Cigars
Design: Jan Horcik
@heavyweight_type

[1] **Zetafonts**
Typeface: Aquawax
Design: Francesco Canovaro
@zetafonts

[2] **Ultra Types**
Typeface: UT Barrel
Design: Wete
@ultra_types

[3] **Optimo**
Typeface: Practice Pan
Design: François Rappo
@optimofoundry

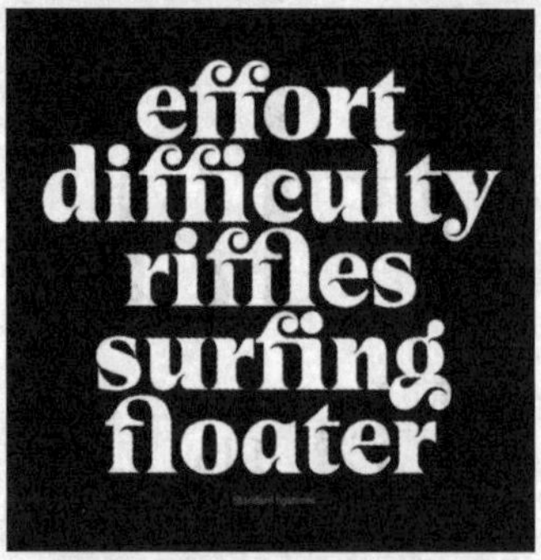

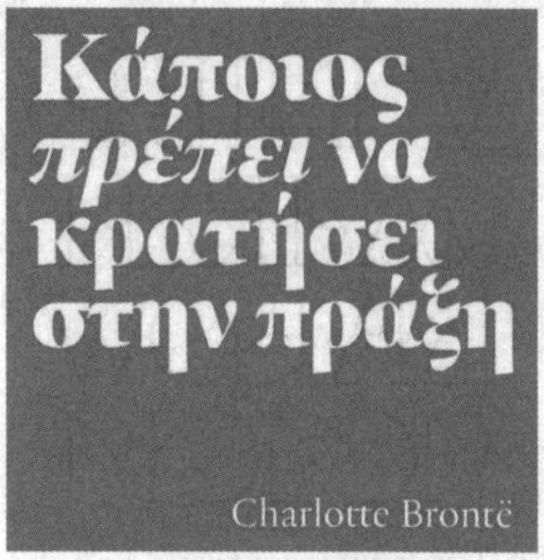

[1] **The Designers Foundry**
Typeface: Skanaus
Design: Daniel Gremme
@tdfoundry

[2] **F37**
Typeface: F37 Bella
Design: Rick Banks
@f37foundry

[3] **Positype**
Typeface: Crave Aura
Design: Neil Summerour
@positype

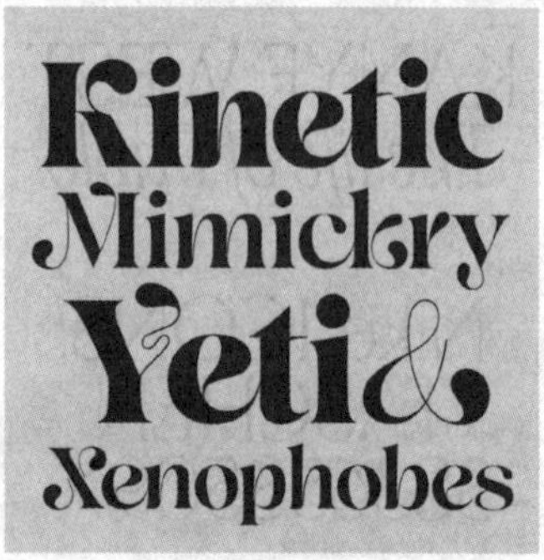

[2] **Road to Venice Type**
Typeface: No. 43
Design: Road to Venice Type
@r.v.type

[3] **R-Typography**
Typeface: Rizoma S
Design: R-Typography
@r_typography

180 pt

(3) Calm Type

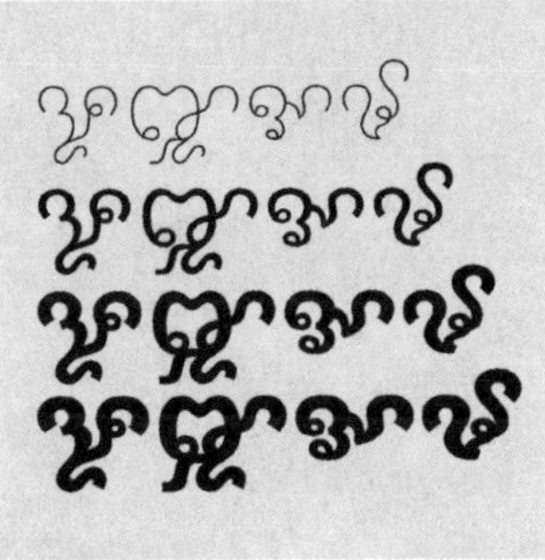

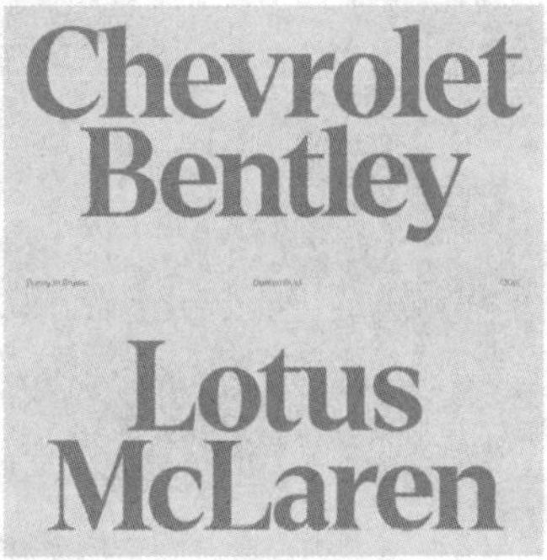

[1] **The Fontpad**
Typeface: Noto
Design: Mark Frömberg, Ben Mitchell
@thefontpad

[2] **Peregrin Studio**
Typeface: Denton
Design: Peregrin Studio
@peregrinstudio

[3] **Or Type**
Typeface: Lemmen
Design: Or Type
@ortype

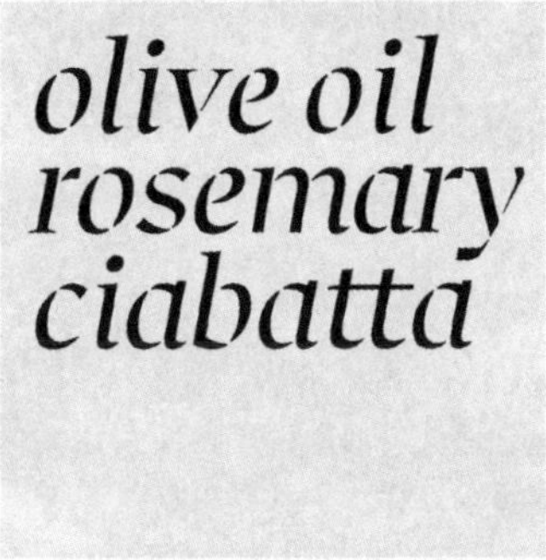

[1] **DDOTT**
Typeface: Nouveau
Design: Dominik Thieme
@ddott_type

[2] **Faire Type**
Typeface: Toast Stencil
Design: Faire Type
@fairetype

[3] **Jakob Fangmeier**
Typeface: Stein
Design: Jakob Fangmeier
@jakobfangmeier

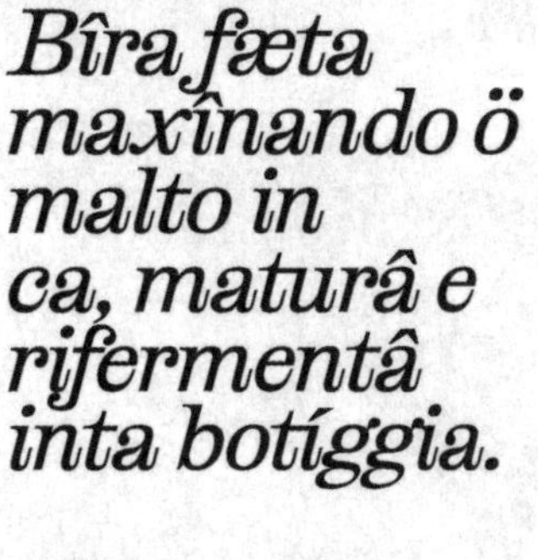

[1] **Pangram Pangram**
Typeface: Editorial Old
Design: Mat Desjardins, Fred Wiltshire, Francesca Bolognini
@pangram.pangram

[2] **Diorama Type Partners**
Typeface: Diorama Gothic
Design: Diorama Type Partners
@dioramatypepartners

[3] **Scott Vander Zee**
Typeface: Scotch Genovese
Design: Scott Vander Zee
@s.vd.z

[1] **Range Left**
Typeface: Bakerie
Design: Clément Cases
@rangeleftshop

[2] **Nouvelle Noire**
Typeface: NN Didot Modern
Design: Arnaud Chemin
@nouvellenoire_ch

[3] **Typotheque**
Typeface: November
Design: Typotheque
@typotheque

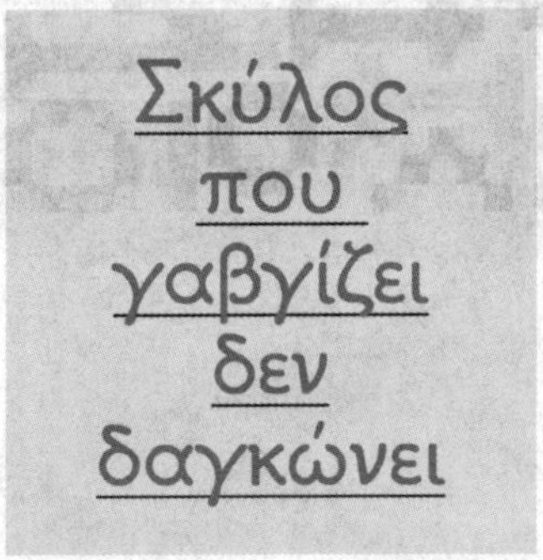

[1] **Parachute**
Typeface: Alfabeta
Design: Panos Vassiliou, Victor Scholderer
@parachutetypefoundry

[2] **Margot Lévêque Type Foundry**
Typeface: Romie
Design: Margot Lévêque
@margot.leveque

(3) Calm Type

Zin Nağao
Typeface: ZNVT17
Design: Zin Nağao
—
@zinnağao

Astrae Studio
Typeface: Chateau
Design: Astrae Studio
—
@astrae.studio

Florian Karsten Typefaces
Typeface: FK Raster
Design: Květoslav Bartoš
—
@floriankarsten

182 pt (4) Experimental Type

Güggeli is wild and YOU have to tame her. It's more about bringing something chaotic into a nice balance. You are challenged.

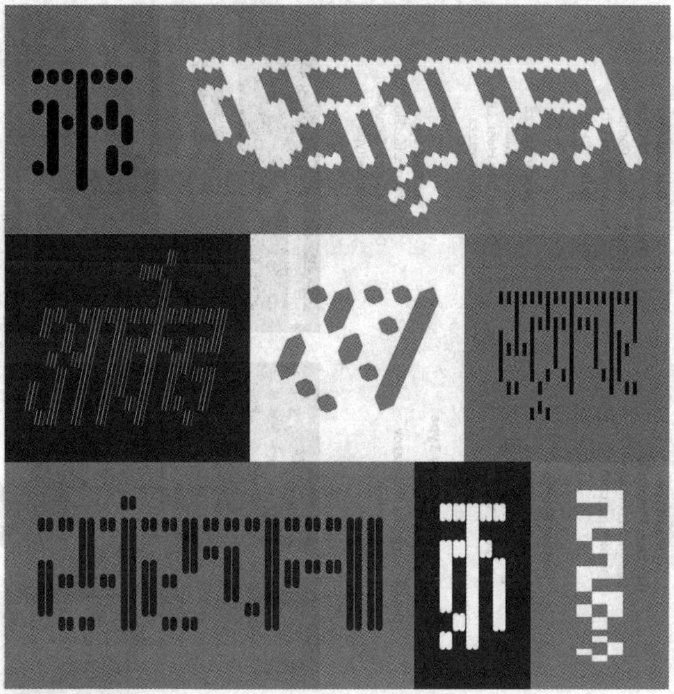

SFB Fonts
Typeface: Güggeli
Design: Studio Fabio Biesel
—
@sfbfonts

Universal Thirst
Typeface: Splinter
Design: Namrata Goyal, Frederik Berlaen, Universal Thirst
—
@universalthirst

Diorama Type Partners
Typeface: Lezen
Design: Diorama Type Partners
—
@dioramatypepartners

[1] **Due Studio**
Typeface: Analo Grotask
Design: Alessio Pompadura, Massimiliano Vitti
@due_std

[2] **Quicknap.zzz**
Typeface: Mikon
Design: Dae Huen Lee
@quicknap.zzz

[3] **No Bad Type**
Typeface: Cigarillo
Design: Juuso Salakka
@nobadtype

[1] **Maxitype**
Typeface: Metaballs
Design: Maximage
@maxitype_com

[2] **Due Studio**
Typeface: Neue Brücke
Design: Massimiliano Vitti
@due_std

[3] **Seryozha Rasskazov Type**
Typeface: Didonist
Design: Seryozha Rasskazov
@typetention

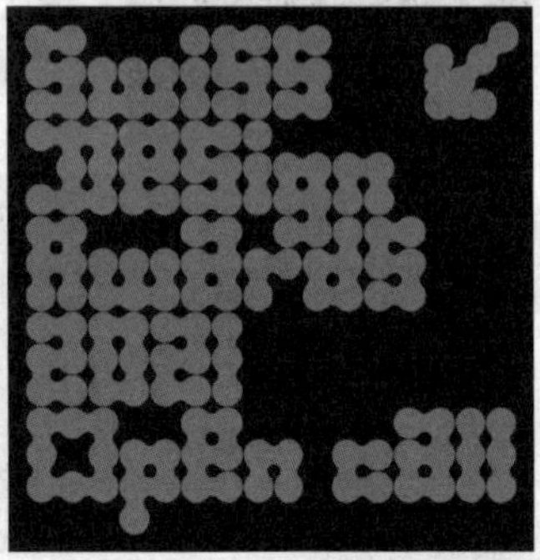

[1] **TEKIO**
Typeface: Ursa Mono
Design: Jan Estrada-Osmycki
@tekio.work

[2] **Skrr Type**
Typeface: Carbone
Design: Eugénie Boron
@skrr_type

[3] **Altiplano**
Typeface: Monaako
Design: Altiplano
@altiplano_typefaces

[1] **Typografische**
Typeface: Alpha
Design: Typografische
@typografische

[2] **Gruppo Due**
Typeface: G2 Ciao
Design: Massimiliano Audretsch
@gruppo.due

[3] **DDOTT**
Typeface: Aether
Design: Dominik Thieme
@ddott_type

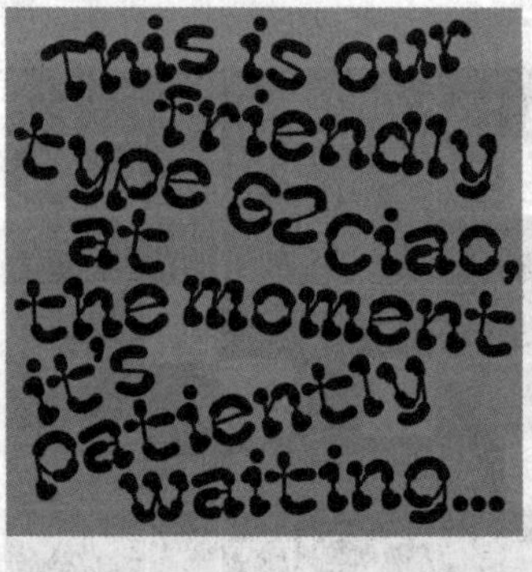

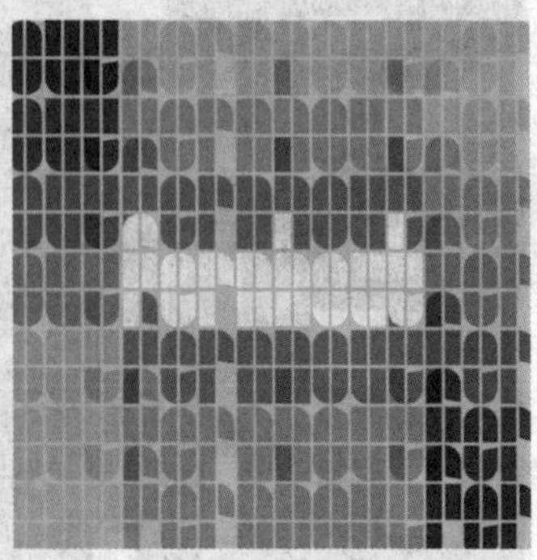

[2] **W Type Foundry**
Typeface: Neo-Geo
Design: David Súid
@wtypefoundry

[3] **The Foundry Types**
Typeface: Fernhout
Design: The Foundry Types
@thefoundrytypes

184 pt

(4) Experimental Type

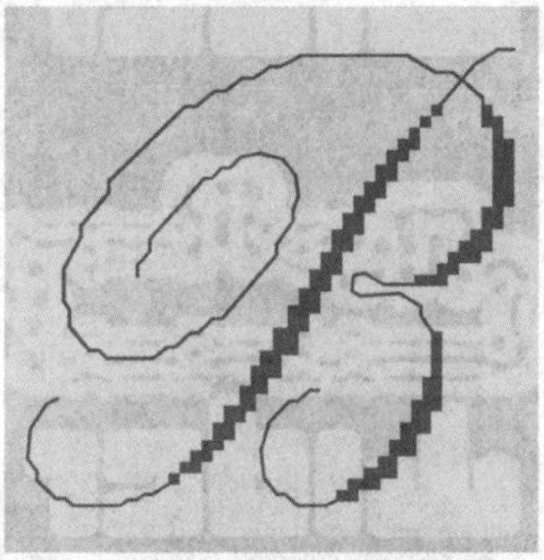

[1] **Biliktu Foundry**
Typeface: Erkin
Design: Murathan Biliktü
@biliktufoundry

[2] **SUPERCONTINENTE**
Typeface: Rym
Design: Fabiola Mejía
@supercontinente_

[3] **Ryan Buğden**
Typeface: Ichigo
Design: Ryan Buğden
@ryanbugden

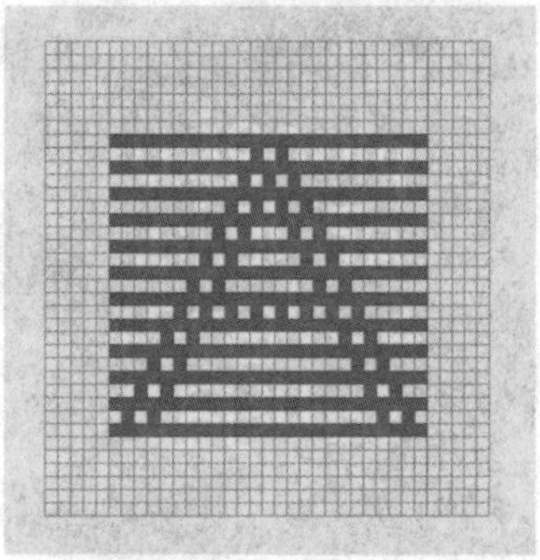

[1] **Daytona Mess**
Typeface: Baleze
Design: Anne-Dauphine Borione
@daytonamess.otf

[2] **CAST Type Foundry**
Typeface: Gil Modern
Design: Valerio Monopoli
@castfoundry

[3] **Nouvelle Noire**
Typeface: Knit Grotesk
Design: Rüdiger Schlömer
@nouvellenoire_ch

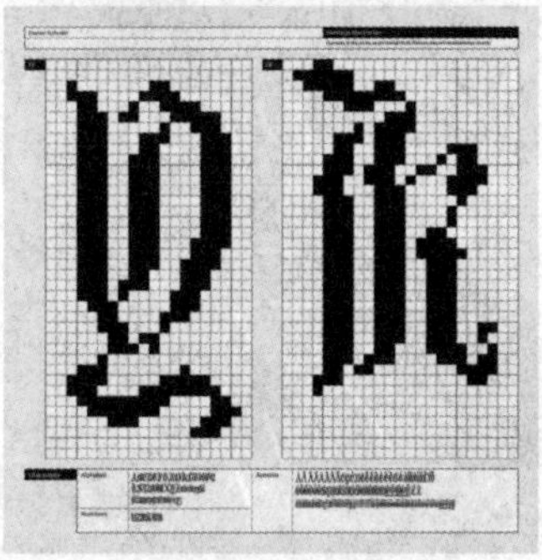

[1] **Collettivo**
Typeface: Sinistre
Design: Jules Durand
@collettivo

[2] **A2Z Foundry**
Typeface: Heritage
Design: Daniel Schrïer
@azzfoundry

[3] **Eugénie Boron**
Typeface: Echo
Design: Eugénie Boron
@eugenieboron

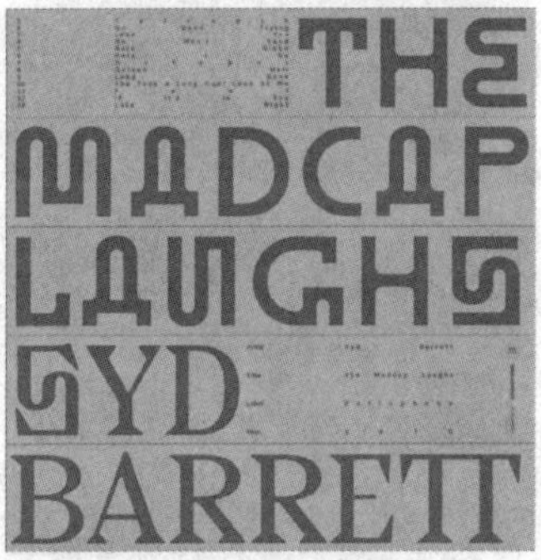

[1] **La Bolde Vita**
Typeface: BHV
Design: Fabian Dornhecker
@laboldevita

[2] **Studio Feixen**
Typeface: Sans Edgy
Design: Felix Pfäffli, Robin Eberwein
@studiofeixen

[3] **Elias Hanzer**
Typeface: Medieval Edition
Design: Elias Hanzer
@elias_hanzer

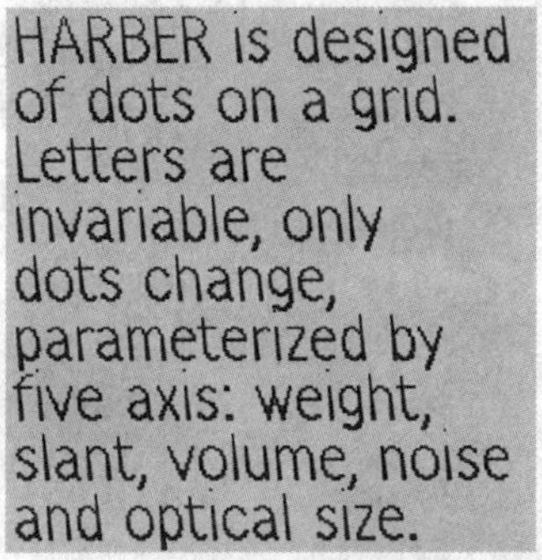

[1] **bb-bureau**
Typeface: Harber
Design: Benoît Bodhuin
@benoitbodhuin

[2] **Shriftovik**
Typeface: Sofuto
Design: Valeria Shumova
@shriftovik

[1] **Metis Foundry**
Typeface: Molten
Design: Simon Bent
@metis_foundry

[2] **Plain Form**
Typeface: Keyframe
Design: Lucas Descroix
@plain_form

[3] **Bloom Type Foundry**
Typeface: Knopp
Design: Bloom Type Foundry
@bloomtypefoundry

[1] **Le Laïc: Type Foundry**
Typeface: Nieuk
Design: Maciej Połczyński
@laictype

[2] **Maria Montes**
Typeface: Green Fairy
Design: Maria Montes
@lamariamontes

[3] **Tightype**
Typeface: Macan
Design: Tightype
@tightype

[1] **Lettersoup**
Typeface: Milka
Design: Lettersoup
@lettersoup_type_foundry

[2] **Swiss Typefaces**
Typeface: Raskal Oner
Design: Emmanuel Rey / Swiss Typefaces
@swisstypefaces

[3] **Monomodular Type Foundry**
Typeface: Wim Mono
Design: Lena Weber
@lenaweber404

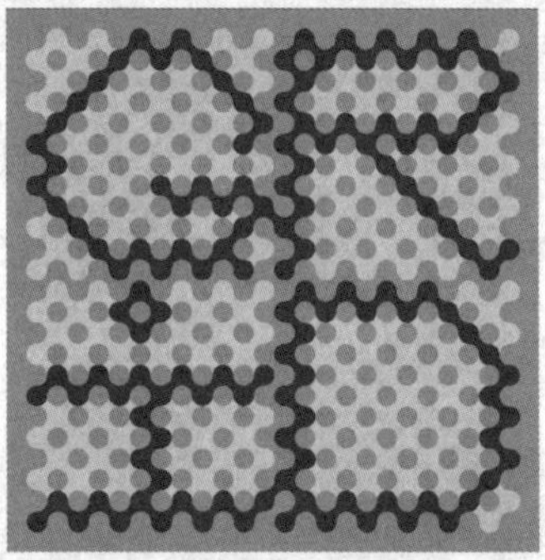

[1] **Outline Online**
Typeface: OO Theran
Design: Laura Csocsán
@cs_laura @outline___online

[2] **Stefanie Vogl**
Typeface: Autark
Design: Stefanie Vogl
@omfdofficial

[3] **New Letters**
Typeface: Laif
Design: Armin Brenner, Markus John
@new_letters

[2] **SUPERCONTINENTE**
Typeface: Segueta
Design: Fabiola Mejía
@supercontinente_

[3] **Typeland**
Typeface: Amifer Folio
Design: Typeland
@typeland

180 pt (4) Experimental Type

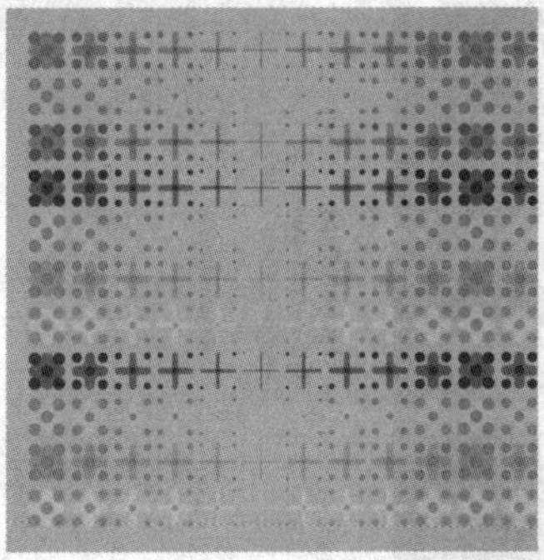

[1] **Fontwerk**
Typeface: Hamster
Design: Joe Stitzlein
@fontwerk

[2] **Nathan Laurent**
Typeface: Frames
Design: Nathan Laurent
@nathan.laurent.xyz

[3] **AllCaps**
Typeface: Visual
Design: Jan Novák
@allcapstype

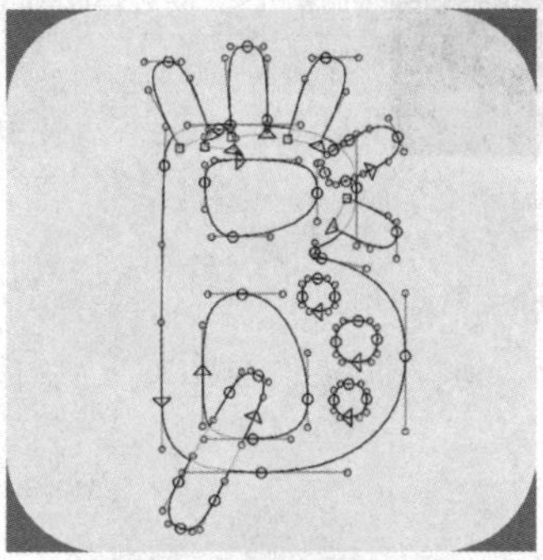

[1] **Vectro Type Foundry**
Typeface: Kablammo
Design: Travis Kochel
@vectrotype

[2] **Skrr Type**
Typeface: Kaolin
Design: Agathe Challeton
@skrr_type

[3] **PFA Typefaces**
Typeface: Knochen
Design: Martin Aleith
@pfa_typefaces

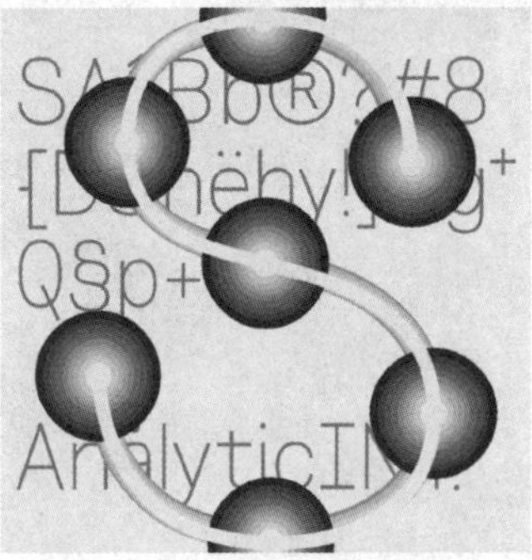

[1] **Why Foundry**
Typeface: Wf Node
Design: Marinus Klinksik
@why.foundry

[2] **Power Type**
Typeface: OffBit
Design: Teguh Arief
@powertypefoundry

[3] **Eliott Grunewald**
Typeface: Harry & Rigaldon
Design: Eliott Grunewald
@eliottgrunewald

[1] **Source Type**
Typeface: Reform
Design: Laurenz Brunner, Fabio Menet
@sourcetype

[2] **Saad Khurshid**
Typeface: BABA
Design: Saad Khurshid
@bysaadkhurshid

[3] **Sandrine Nugue**
Typeface: Skeleton
Design: Sandrine Nugue
@sndrnng

[1] **Metis Foundry**
Typeface: Geometer
Design: Simon Bent
@metis_foundry

[2] **Boafff**
Typeface: Rabbit Hole
Design: boafff
@boafff

76,126 miles
300 %
[DOT–GAIN]
1.499 $
85 × 55 mm

{أ} بتيث «ج»
(ذنز/٢# شط)
+٧٤ فمقغ ؟
كهه: لا ٠ ٨ ئ

Heheh Type
Typeface: Gawhar
Design: Heheh Type
—
@hehehtype

David Einwaller
Typeface: Bolivia
Design: David Einwaller
—
@d.einw

MCKL
Typeface: Owners
Design: Jeremy Mickel
—
@mckltype

plastic
typo
graphy

188 pt (5) Technical Type

BIOMETRICAL ANALYSIS
IAAF WORLD CHAMPIONSHIPS IN ATHLETICS BERLIN, 15.–23.08.2009

NAME	ROUND	WIND	RT	T 20M	T 40M	T 60M	T 80M	T 100M
Usain Bolt	Fi	0,9	0,146	2,89	4,64	6,31	7,92	9,58
Usain Bolt	SF 1	0,2	0,135	2,89	4,68	6,41	8,11	9,89
Tyson Gay	Fi	0,9	0,144	2,92	4,70	6,39	8,02	9,71
Tyson Gay	SF 2	-0,2	0,143	2,99	4,80	6,54	8,21	9,93
Asafa Powell	Fi	0,9	0,134	2,91	4,71	6,42	8,10	9,84
Asafa Powell	SF 2	-0,2	0,133	2,92	4,73	6,47	8,17	9,95
Daniel Bailey	Fi	0,9	0,129	2,92	4,73	6,48	8,18	9,93
Daniel Bailey	SF 1	0,2	0,135	2,93	4,74	6,49	8,19	9,96
Richard Thompson	Fi	0,9	0,119	2,90	4,71	6,45	8,17	9,93
Richard Thompson	SF 2	-0,2	0,132	2,92	4,74	6,51	8,22	9,98
Dwain Chambers	Fi	0,9	0,123	2,93	4,75	6,50	8,22	10,00
Dwain Chambers	SF 2	-0,2	0,182	2,96	4,79	6,55	8,26	10,04
Marc Burns	Fi	0,9	0,165	2,94	4,76	6,52	8,24	10,00
Marc Burns	SF 1	0,2	0,159	2,95	4,76	6,52	8,23	10,01
Darvis Patton	Fi	0,9	0,149	2,96	4,85	6,65	8,42	10,34
Darvis Patton	SF 1	0,2	0,152	2,96	4,78	6,51	8,21	9,98
Michael Rodgers	SF 1	0,2	0,154	2,95	4,79	6,55	8,27	10,04
Monzavous Edwards	SF 2	-0,2	0,146	3,01	4,89	6,67	8,38	10,14
Michael Frater	SF 2	-0,2	0,153	2,97	4,81	6,59	8,31	10,14
Martial Mbandjock	SF 1	0,2	0,138	3,06	4,92	6,69	8,41	10,18
Gerald Phiri	SF 2	-0,2	0,143	2,93	4,79	6,59	8,36	10,19
Jaysuma Saidy Ndure	SF 1	0,2	0,143	2,96	4,81	6,60	8,36	10,20
Naoki Tsukahara	SF 2	-0,2	0,152	2,98	4,86	6,67	8,44	10,25

It'll End in Tears
CAD 411—
October 1984

Coil Light
25.09.2020
v. 1.00

Kimera
Typeface: Apparat
Design: Michael Clasen, Marcel Saidov
—
@kimeracorp

Brownfox
Typeface: Coil
Design: Vyacheslav Kirilenko, Gayaneh Bağdasaryan
—
@brownfoxtype

№Δëŭŕæl
¼ŊĒŤẄÕ
Ŕ…“K”™±
9€µⓅM∞

ECAL Typefaces
Typeface: Neureal
Design: Laura Csocsán
—
@ecalmatd

[1] **Mass-Driver**
Typeface: MD IO
Design: Rutherford Craze
mass-driver.com

[2] **Process Type Foundry**
Typeface: Sculpin
Design: Eric Olson
@process_type

[3] **Razzia Type**
Typeface: RT Alias
Design: Mirco Schiavone, RazziaType
@razziatype

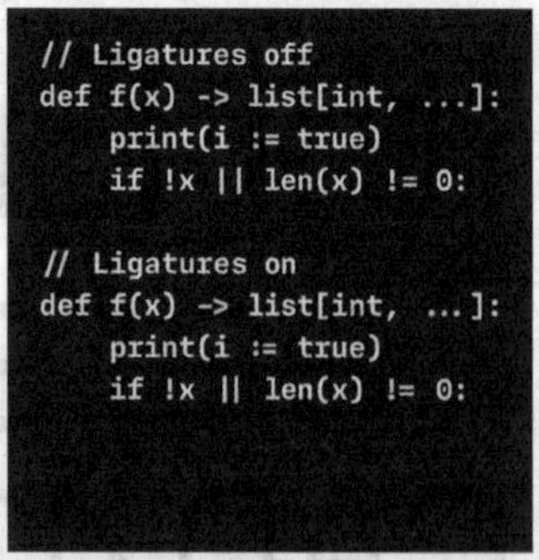

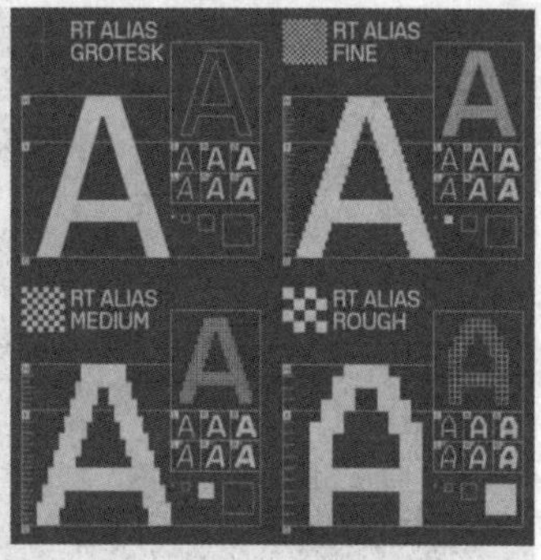

[1] **Sociotype**
Typeface: Gestura
Design: Sociotype
@sociotype.foundry

[2] **Dinamo Typefaces**
Typeface: ROM
Design: Seb McLauchlan
@abcdinamo

[3] **Dalton Maag**
Typeface: Aktiv Grotesk
Design: Dalton Maag
@dalton.maag

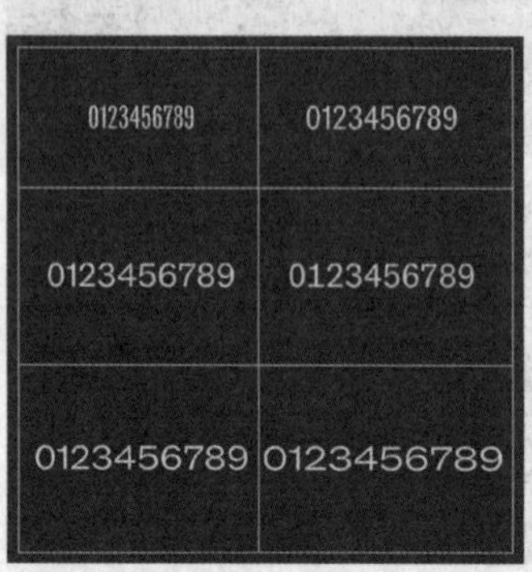

[1] **Typeji**
Typeface: Looklyn
Design: Tien-Min Liao
@typeji

[2] **Jo Malinis**
Typeface: Tuka
Design: Jo Malinis
@aniciaclean

[3] **Dirtyline Studio**
Typeface: Neue Metana Mono
Design: Dirtyline Studio
@dirtyline.std

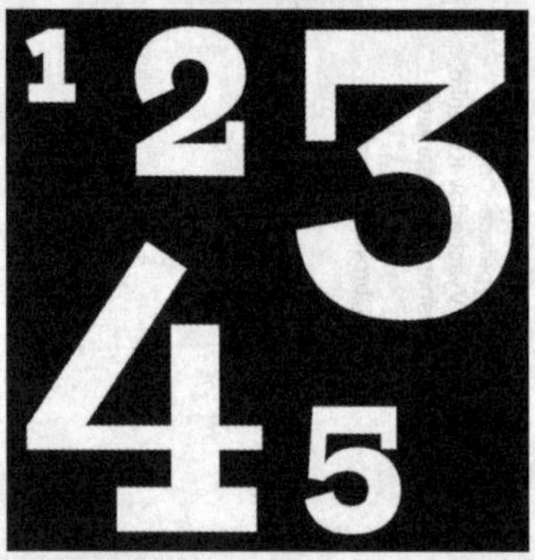

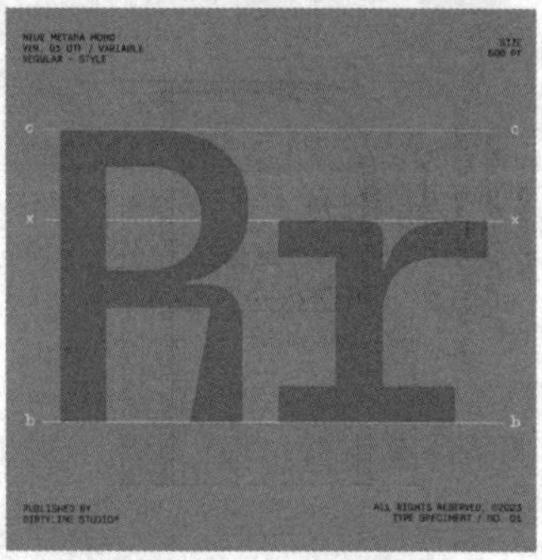

[1] **Jan Fromm**
Typeface: Capito
Design: Jan Fromm
@jan__fromm

[2] **Feedtype**
Typeface: Feed Sans
Design: Feed
@feedtype

[3] **APFEL Type Foundry**
Typeface: Friedel
Design: APFEL Type Foundry
@apfeltypefoundry

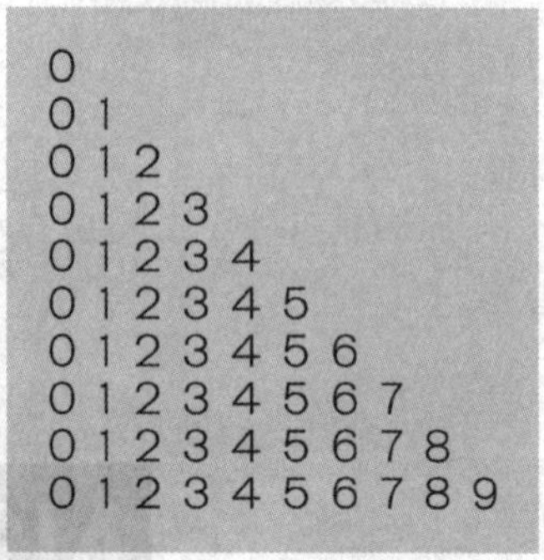

[2] **Family Type**
Typeface: Athletics
Design: Family Type
@familytype.co

[3] **Modern Type**
Typeface: Everyday Sans
Design: Dalton Bruyns
@moderntype_

pt (5) Technical Type

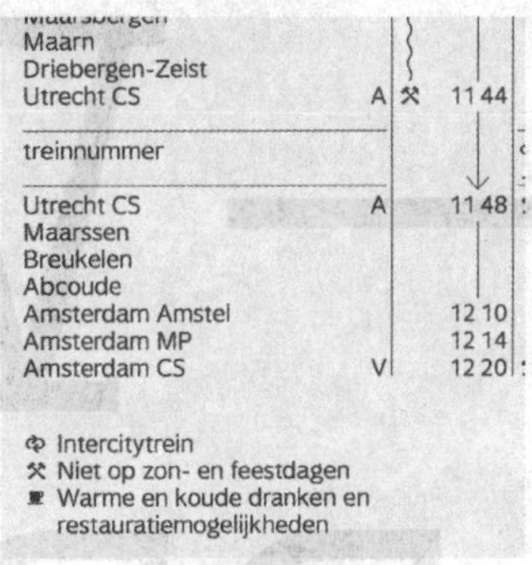

[1] **DDOTT**
Typeface: Aether
Design: Dominik Thieme
@ddott_type

[2] **Kanon Foundry**
Typeface: Diagol Grotesk
Design: Tor Weibull
@kanon_foundry

[3] **WiseType**
Typeface: Saltburn
Design: Jacob Wise
@wise_type

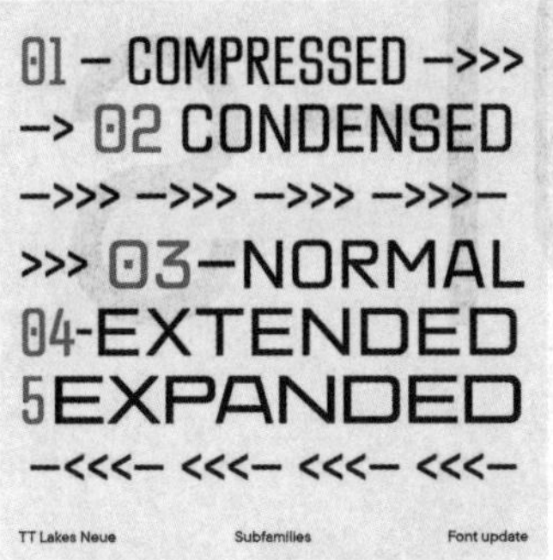

[1] **Swiss Typefaces**
Typeface: Euclid Mono
Design: Emmanuel Rey,
Quentin Schmerber / Swiss Typefaces
@swisstypefaces

[2] **TypeType**
Typeface: TT Lakes Neue
Design: TypeType
@typetype.foundry

[3] **Heavyweight**
Typeface: Pacific & Atlantic
Design: Jan Horcik
@heavyweight_type

[1] **Pangram Pangram**
Typeface: Neue Monteral
Design: Mat Desjardins
@pangram.pangram

[2] **NaN**
Typeface: Tragedy Mono
Design: Jean-Baptiste Morizot,
Léon Hugues, Daria Cohen
@nan_xyz

[3] **ECAL Typefaces**
Typeface: Mule
Design: Samira Schneuwly
@ecal_ch

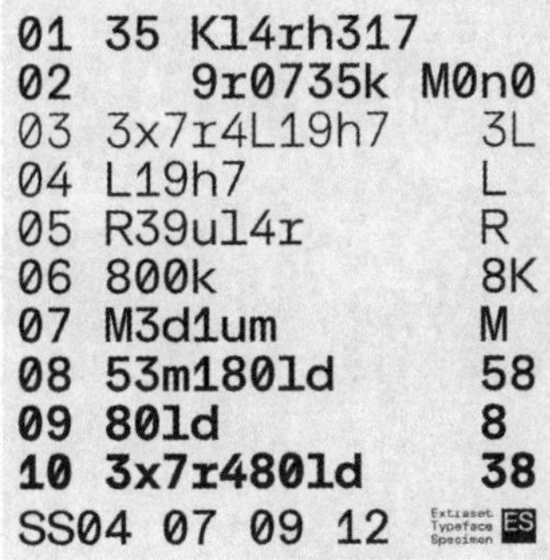

[1] **Extraset**
Typeface: ES Klarheit Grotesk
Design: Extraset
@extraset.ch

[2] **Florian Karsten Typefaces**
Typeface: FK Grotesk
Design: Florian Karsten
@floriankarsten

[3] **Type Supply**
Typeface: Epoxy
Design: Tal Leming
@typesupply

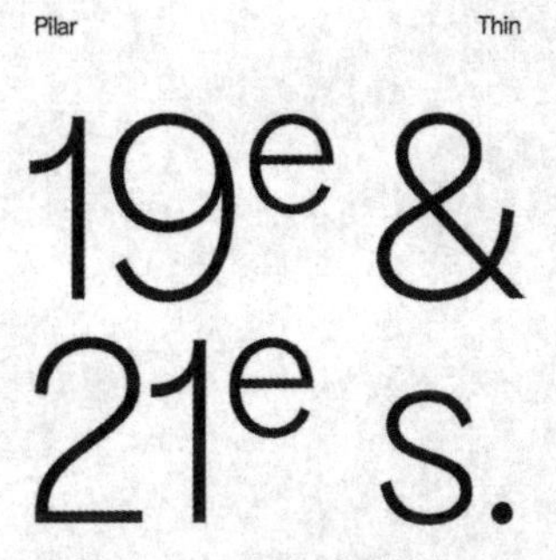

[1] **Tightype**
Typeface: Habitas
Design: Fabian Fohrer, Léa Bruneau
@tightype

[2] **Fuerte Type**
Typeface: Pilar
Design: Fermin Guerrero
@fuertetype

—VI—

PDFs

192pt

PDF specimens, initially designed for digital screens, are a vital part of modern typographic documentation. Here, we've printed these digital specimens to preserve them as tangible records of contemporary type presentation methods. This chapter highlights how designers embrace the PDF format to creatively communicate a font's character, features, and potential—now immortalized on paper. But not just any paper—we chose something special for these treasures, adding depth and a touch of transparency to elevate their presentation.

Diorama Type Partners
Typeface: Bet
Design: Guillaume Sbalchiero
—
dioramatypepartners.com @dioramatypepartners

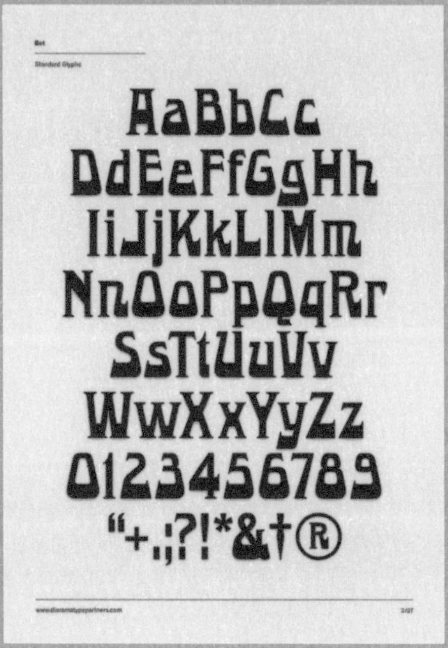

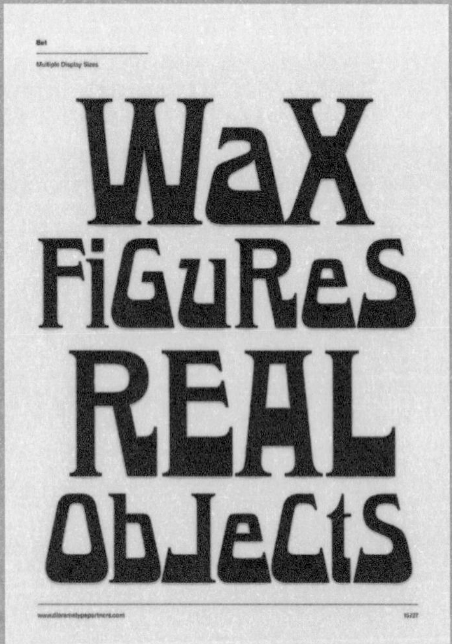

194 pt.

Support Independent Type II

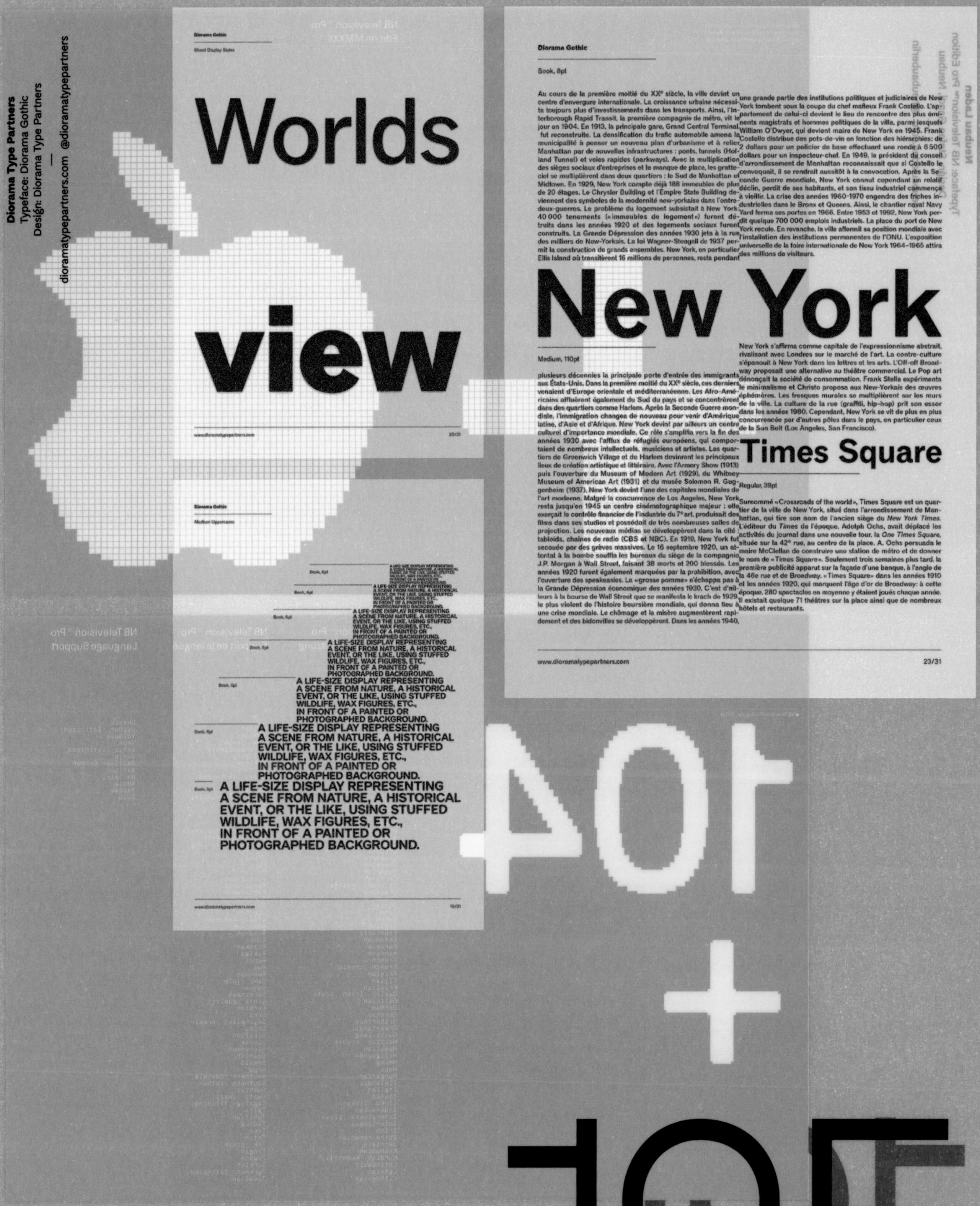

Diorama Type Partners
Typeface: Diorama Gothic
Design: Diorama Type Partners
—
dioramatypepartners.com @dioramatypepartners

Neubau Laden
Typeface: NB Television™ Pro Edition
Design: Stefan Gandl, Neubau
—
neubauladen.com @neubauberlin

Jung-Lee Type Foundry
Typeface: Pirelli
Design: Jungmyung Lee
—
j-ltf.com @jung_lee_type_foundry

Pirelli Thin
Pirelli Thin Italic
Pirelli Medium
Pirelli Medium Italic

: COMPACT
written by Maurice Roche
Molluc Oblique

You shall be made sleepless even as you are left sightless. While you're penetrating the darkness, you'll penetrate into the night, getting in deeper and deeper, your already failing memory growing proportionally weaker as—at the end of a long lethargy—you become conscious of your condition.

(How will you tell day from night?) [1]

You'll slowly

turn your head

to the left ***to the right***
before letting your neck fall back on the damp pillow. The frozen contract of the pillowcase will make you shiver. you'll touch your face, slowly you'll feel it (a presence!); and that object (which?) that—stretching out your arm—you'll displace on the bedside table to your right while leaving the nocturnal landscape unchanged [2]

And then you'll abandon this open night for a **closed** one: **softly**—you'll pull it toward you, you'll attract it to you—you'll lower your eyelids to reduce it to a **small** night that belongs to you (where you'll take refuge, hoping to retrieve the memory of
, and the trail of a day-dream that came to disturb the **endless** shadow...) with **closed** eyes you'll persist in compressing the darkness; with all your might, face in hands palms cupped over eye-sockets... [3]

[1] Pirelli Thin 11 pt
[2] Pirelli Medium 17 pt
Pirelli Medium Italic
[3] Pirelli Thin 21 pt
Pirelli Medium
Jungka Bold

PIRELLI THIN
Impact Nieuw 2018 Italic

AÁĂÂÄÀĀĄÅÃÆBCĆČÇDĐĎĐ
EÉĚÊËĖÈĒĘFGĞHIÍÎÏĪĮJ
KĶLĹĽĻŁMNŃŇŅŊÑ
OÓÔÖÒŐŌØÕŒPÞQRŔŘŖ
SŚŠŞTŦŤŢUÚÛÜÙŰŪŲŮ
VWẂŴẄẀXYÝŶŸỲZŹŽŻ
aáăâäàāąåãæbcćčçdďđeéěêëėèēę
fgģhiıíîïìīįjkķlĺľļłmnńňņŋñ
oóôöòőōøõœpþqrŕřŗsśšşß
tŧťţuúûüùűūųůvwẃŵẅẁx
yýŷÿỳzźžż
0123456789*
·•:;,...!¡#.?¿"';//_{}[]()—–-«»‹›„"""',
/.¢¤$€£¥≈~÷ø=>≥∞∫<≤¬
−×≠%‰+±∏√∑
◊|¦@&¶©®§™°^†‡
↑↗→↘↓↙↖↔↕

Pirelli Medium
Pirelle Medium

AÁĂÂÄÀĀĄÅÃÆBCĆČÇDĐĎĐ
EÉĚÊËĖÈĒĘFGĞHIÍÎÏĪĮJ
KĶLĹĽĻŁMNŃŇŅŊÑ
OÓÔÖÒŐŌØÕŒPÞQRŔŘŖ
SŚŠŞTŦŤŢUÚÛÜÙŰŪŲŮ
VWẂŴẄẀXYÝŶŸỲZŹŽŻ
aáăâäàāąåãæbcćčçdďđeéěêëėèēę
fgģhiıíîïìīįjkķlĺľļłmnńňņŋñ
oóôöòőōøõœpþqrŕřŗsśšşß
tŧťţuúûüùűūųůvwẃŵẅẁx
yýŷÿỳzźžż
0123456789*
·•:;,...!¡#.?¿"';//_{}[]()—–-«»‹›„"""',
/.¢¤$€£¥≈~÷ø=>≥∞∫<≤¬
−×≠%‰+±∏√∑
◊|¦@&¶©®§™°^†‡
↑↗→↘↓↙↖↔↕

Pirelli Bold
Joseleen Oblique

AÁĂÂÄÀĀĄÅÃÆBCĆČÇDĐĎĐ
EÉĚÊËĖÈĒĘFGĞHIÍÎÏĪĮJ
KĶLĹĽĻŁMNŃŇŅŊÑ
OÓÔÖÒŐŌØÕŒPÞQRŔŘŖ
SŚŠŞTŦŤŢUÚÛÜÙŰŪŲŮ
VWẂŴẄẀXYÝŶŸỲZŹŽŻ
aáăâäàāąåãæ
bcćčçdďđeéěêëėèēę
fgģhiıíîïìīįjkķlĺľļłmnńňņŋñ
oóôöòőōøõœpþqrŕřŗsśšşß
tŧťţuúûüùűūųůvwẃŵẅẁx
yýŷÿỳzźžż
0123456789*
·•:;,...!¡#.?¿"';//_{}[]()—–-«»‹›„"""',
/.¢¤$€£¥≈~÷ø=>≥∞∫<≤¬
−×≠%‰+±∏√∑
◊|¦@&¶©®§™°^†‡
↑↗→↘↓↙↖↔↕

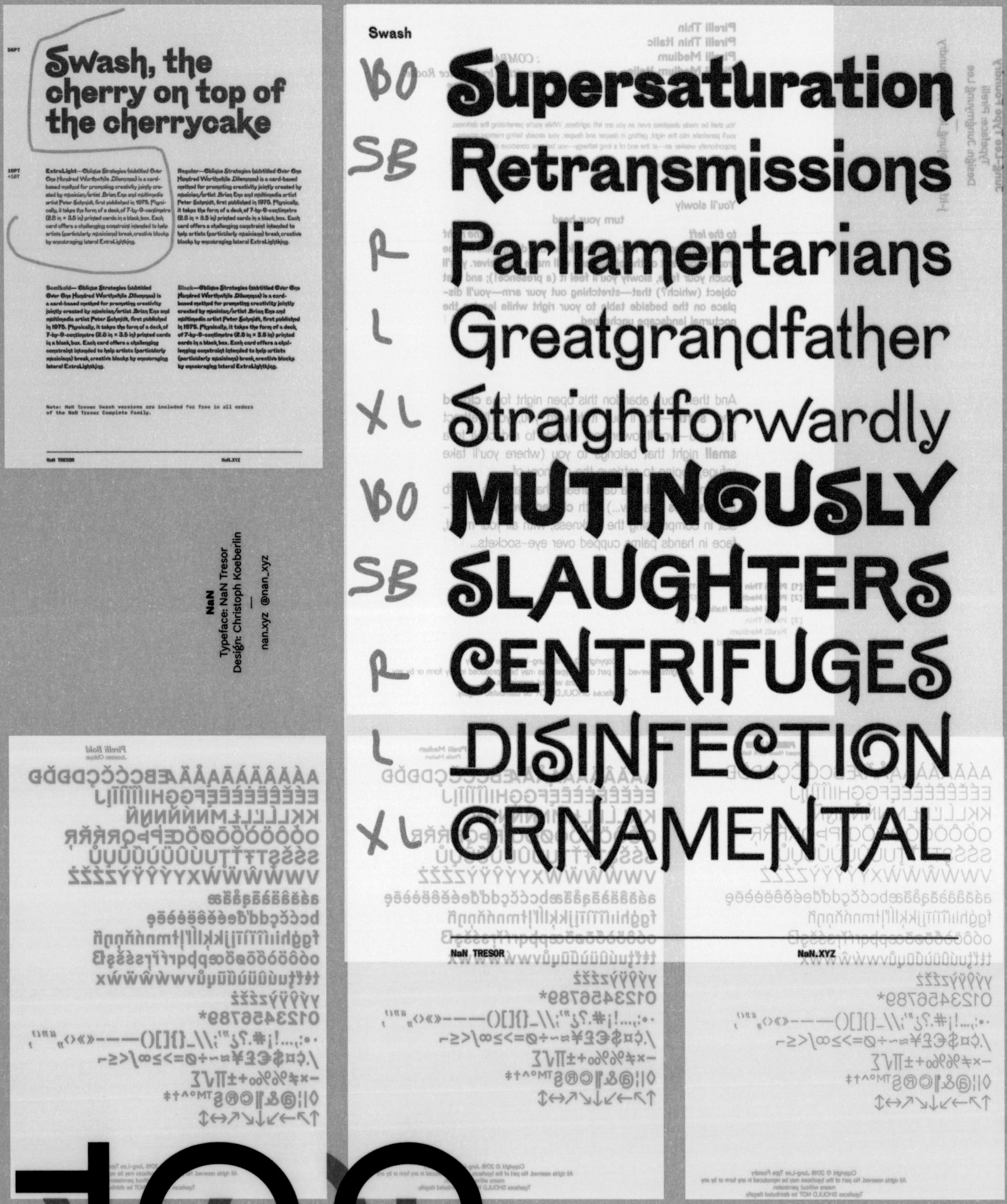

NaN
Typeface: NaN Tresor
Design: Christoph Koeberlin
—
nan.xyz @nan_xyz

NaN Tresor S ExtraLight

84PT -25T

Revivals

32PT -15T

The resurrection of the dead is a standard eschatological belief in the Abrahamic

18PT

In the next decade Howard Ruff warned about socio-economic collapse in his 1974 book Famine and Survival in America. Ruff's book was published during a period of rampant inflation in the wake of the 1973 oil crisis. Most of the elements of survivalism can be found there,

12PT +10T

In the next decade Howard Ruff warned about socio-economic collapse in his 1974 book Famine and Survival in America. Ruff's book was published during a period of rampant inflation in the wake of the 1973 oil crisis. Most of the elements of survivalism can be found there, including advice on food storage. The book championed the claim that precious metals, such as gold and silver, have an intrinsic worth that makes them more usable in the event of a socioeconomic collapse than fiat currency. Ruff later published milder variations

NaN TRESOR NaN.XYZ

NaN Tresor M Bold

84PT -10T

Revivals

32PT -5T

The resurrection of the dead is a standard eschatological belief in

18PT

In the next decade Howard Ruff warned about socio-economic collapse in his 1974 book Famine and Survival in America. Ruff's book was published during a period of rampant inflation in the wake of the 1973 oil crisis. Most of the elements of survivalism can be found there,

12PT +10T

In the next decade Howard Ruff warned about socio-economic collapse in his 1974 book Famine and Survival in America. Ruff's book was published during a period of rampant inflation in the wake of the 1973 oil crisis. Most of the elements of survivalism can be found there, including advice on food storage. The book championed the claim that precious metals, such as gold and silver, have an intrinsic worth that makes them more usable in the event of a socioeconomic collapse than fiat currency. Ruff later published milder varia-

NaN TRESOR NaN.XYZ

NaN Tresor Swash Bold

(Included in all orders of the NaN Tresor Complete Family)

84PT -10T

Revivals

32PT -5T

The resurrection of the dead is a standard eschatological belief in the Abrahamic

18PT

In the next decade Howard Ruff warned about socio-economic collapse in his 1974 book Famine and Survival in America. Ruff's book was published during a period of rampant inflation in the wake of the 1973 oil crisis. Most of the elements of survivalism can be found there, including advice on food storage.

12PT +10T

In the next decade Howard Ruff warned about socio-economic collapse in his 1974 book Famine and Survival in America. Ruff's book was published during a period of rampant inflation in the wake of the 1973 oil crisis. Most of the elements of survivalism can be found there, including advice on food storage. The book championed the claim that precious metals, such as gold and silver, have an intrinsic worth that makes them more usable in the event of a socioeconomic collapse than fiat currency. Ruff later published milder variations of the same themes, such as How to Prosper

NaN TRESOR NaN.XYZ

Opentype Features

Stylistic Set 1: Contrast

Grade Raganque
→ Grade Raganque

Stylistic Set 2: Titelsatz-Umlaute

RÄUME → RÄUME

Swash

Vanytume → Vanytume

Tabular figures

2148 → 2148

Fractions, numerators, denominators

1/2+3/4+H2O
→ ½+¾+H_2O

Ordinals

No. 2a 5o → № 2ª 5º

Case sensitive punctuation

(H-H) → (H-H)

NaN TRESOR NaN.XYZ

Semi-Bold 113 / 50 pts

“The Best View Comes After The Hardest Climb.”

[54] [55]

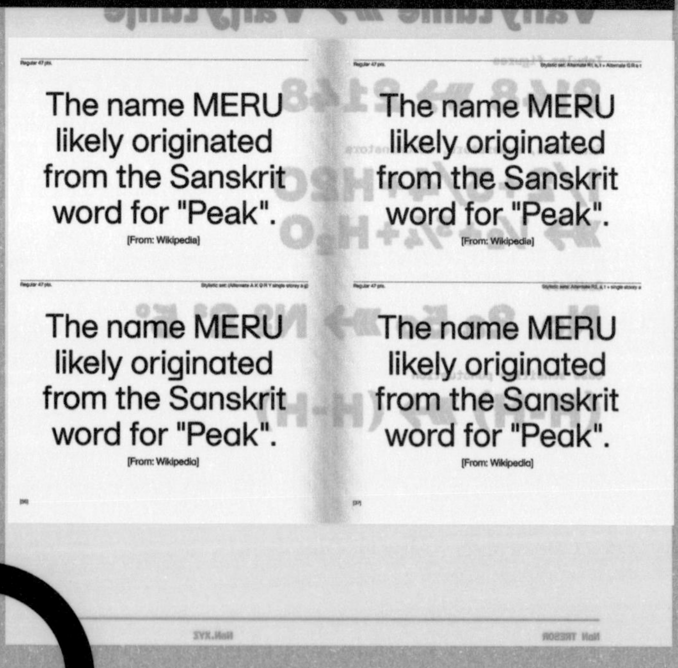

Autograph
Typeface: APK Protocol
Design: Peter Korsman
—
autograph.works @autograph__

200 pt

Support Independent Type II

Uborka
96 pt
4/8
Monospace
Pointed
Editorial
Gurke+Uby
Verboten!
+2.5.1987
Workshops
Hungarumlaut
www.hungarumlaut.com
Hungarumlaut
Typeface: Uborka
Design: Adam Katyi
—
hungarumlaut.com @hungarumlaut

Universal Thirst
Typeface: abacaxi
Design: Gabriel Markan, Gunjan Panchal, Gunnar Vilhjálmsson, Joana Ranito, Hitesh Malaviya, Kalapi Gajjar and Namrata Goyal.
—
universalthirst.com @universalthirst

Universal Thirst युनिवर्सल थर्स्ट family abacaxi

G g a f t ð
G g a f t ð
G g a f t ð
G g a f t ð
G g a f t ð
G g a f t ð
G g a f t ð

Universal Thirst युनिवर्सल थर्स्ट family abacaxi

दिल

200 pt bold

Universal Thirst युनिवर्सल थर्स्ट family abacaxi

extra bold

अपने पिछले पैरों पर खड़े होने पर, एक वयस्क नर ध्रुवीय भालू १० फीट से अधिक लंबा हो सकता है। इसका विशाल आकार इसे प्रभावशाली बल देता है, जो इसके विशाल पंजे के - जोकि १२ इंच चौड़ा हो सकता है - एकल आघात से क्षति का कारण बन सकता है। ये भारी फ़र वाले पैर हिम और बर्फ पर भालू के वजन को वितरित करने में मदद करते हैं, जबकि मोटे, तीखे पंजों का एक सेट अदृष्ट शिकार का इंतजार करते हैं। ध्रुवीय भालू को अक्सर समुद्री बर्फ के किनारे पर दुबकते हुए पाया जाता है, जो सब्र के साथ उनके पसंदीदा भोजन सील का हवा खाने के लिए ऊपर आने का इंतजार करते हैं। लेकिन जब भोजन दुर्लभ हो जाता है, तो भालू अपने ख़ुद के वसा के संचय से, एक दिन में एक किलोग्राम तक जलाकर, बच जाते

When standing on its hind legs, an adult male polar bear may reach more than 10 feet tall. Its enormous size gives it an impressive amount of brute force, able to cause damage with a single swipe of its massive paw - which can stretch 12 inches wide. These heavily furred feet help distribute the bear's weight over the snow and ice, while a set of thick, sharp claws lie in wait for unsuspecting prey. Polar bears are often be found lurking on the edge of the sea ice, patiently waiting for seals - their favourite meal - to come up for air. But when food is sparse, the

15 pt

Heart

Universal Thirst युनिवर्सल थर्स्ट family abacaxi

पीना कोलाडा

120 pt semi bold

Piña Colada

202

pt

Support Independent Type II

Universal Thirst युनिवर्सल थर्स्ट

family	abacaxi
script	devanagari, latin
variable	yes
weights	thin, light, regular, medium, semi bold, **bold**, **extra bold**
year	2023
file types	ttf, otf, woff, woff2

जीवंत Vivid

तेजस Energetic

उज्ज्वल Radiant

कारमेल Caramel

मजबूती **Robustness**

प्रज्वलित **Ablaze**

रसीला **Lush**

Camelot Typefaces
Typeface: Lelo Mono
Design: Katharina Köhler, Camelot Typefaces
—
camelot-typefaces.com @camelot.type

LELO MONO

Bold *Bold Italic*

APPLAUS!
worldwide
Docenić
zeitgemäß
financial
ZRĘCZNOŚĆ
Pop Shop
Sérialité

BoIt
0731

AT THE TIME WARHOL was the master of Business Art, he established *ANDY WARHOL ENTERPRISES* in 1957, and films, Interview magazine, books and TV programmes were to follow but his operation was *small beer* by contemporary standards. Today,

19/29 Katharina Köhler 2021 CA ME LOT

LELO MONO Alternates Thin *Thin Italic*

i l

Stylistic Set 05: Alternate i, l, r, 1

HIRST WURDE IN BRISTOL GEBOREN, wuchs aber in Leeds auf. Sein Vater, ein Automechaniker, verließ die Familie, als er zwölf Jahre alt war Seine Mutter war eine katholische Irin. Von 1986 bis 1989 belegte Damien Hirst am Goldsmiths College in London den Kurs „Fine Arts". Hirst interessierte sich vor allem für diesen Kurs, weil er sich nicht auf eine einzige Kunstrichtung wie Malerei oder Bildhauerei konzentrieren wollte. *IM JAHR 1988 KURATIERTE ER UNTER DEM TITEL FREEZE (GEFRIEREN) eine Ausstellung in einem leerstehenden Lagerhaus im Hafenviertel von London, in der er und andere Künstler ihre Werke ausstellten. Eine weitere von ihm durchgeführte*

DAMIEN STEVEN DAVID BRENNAN est né à Bristol et a grandi à Leeds. Sa mère, Mary Brennan, épouse William Hirst, lorsqu'il a un an, il portera son nom2. Son beau-père était mécanicien et sa mère employée dans l'administration, au Citizens Advice Bureau (en). Ils divorcent

11pt 14pt Stylistic Set 02, 05 Discretionary Ligatures

mixed media
exhibition
high roller
Sensation
Nihilismus

Stylistic Set 05: Alternate i, l, r, 1

That *contemporary art* might be seen in the first instance as *an investment* was soon made evident by the rise of high-return auctions, the most infamous …

Stylistic Set 02, 05 Discretionary Ligatures

5/29 Katharina Köhler 2021 CA ME LOT

AG Font
Typeface: Happiness Sans
Design: AG Font, lo-ol Type Foundry
—
agfont.com @ag.typography.institute

3. OpenType Feature: 스타일 세트1 – 라틴 전용 문장 부호

해피니스 산스는 라틴에 알맞은 위치로 조정된 문장부호를 스타일 세트로 지원합니다. 문장부호 기본 스타일은 한글에 알맞은 위치로 설정되어 있습니다. 스타일 세트 기능을 사용해 문장부호의 위치를 변경할 수 있습니다.

!해피니스,,.? 기본 문장부호 위치

!Happiness,,.? 라틴 전용 문장부호 위치

4. OpenType Feature: 스타일 세트2 – 숫자 모양 변경

해피니스 산스 숫자 중 1, 2, 7은 두 가지, 0은 세 가지 모양으로 디자인하였습니다. 스타일 세트 기능을 사용해 원하는 형태로 사용할 수 있습니다.

127 0 고정너비 숫자 → 127 00

0 가변너비 숫자 → 00

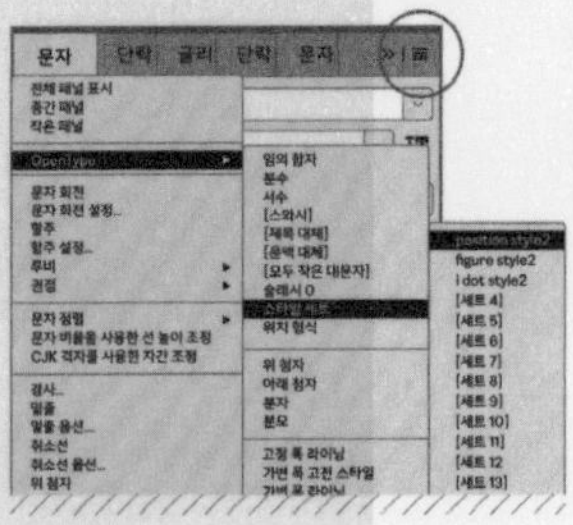

Adobe 인디자인
1. 문자 패널 열기
2. OpenType 선택
3. 스타일 세트 선택
4. position style2 선택

MS 워드
1. 상단 메뉴에서 서식 선택
2. 서식-글꼴
3. 고급 탭
4. 스타일 모음: 1 선택

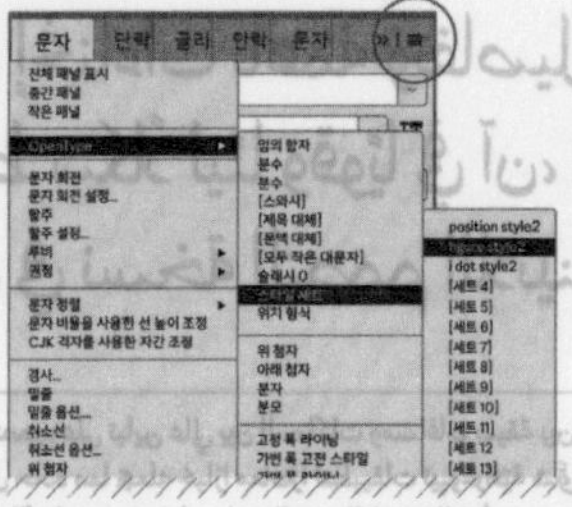

Adobe 인디자인
1. 문자 패널 열기
2. OpenType 선택
3. 스타일 세트 선택
4. figure style2 선택

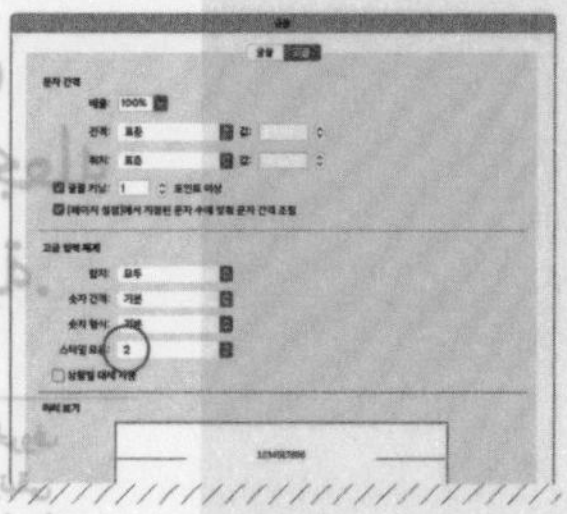

MS 워드
1. 상단 메뉴에서 서식 선택
2. 서식-글꼴
3. 고급 탭
4. 스타일 모음: 2 선택

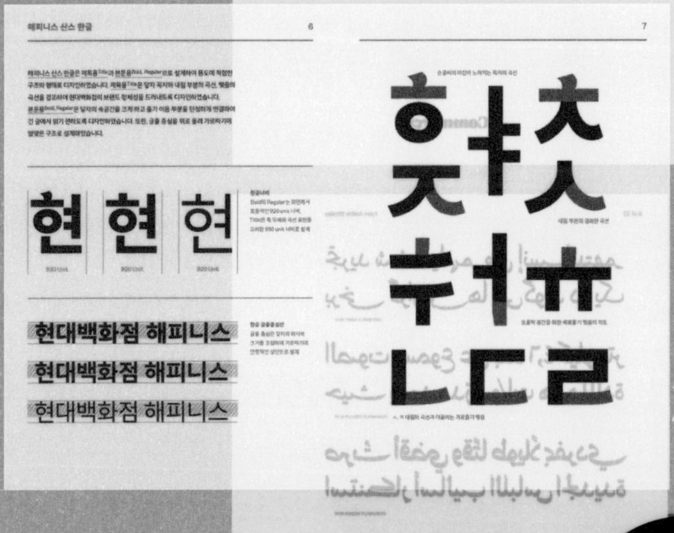

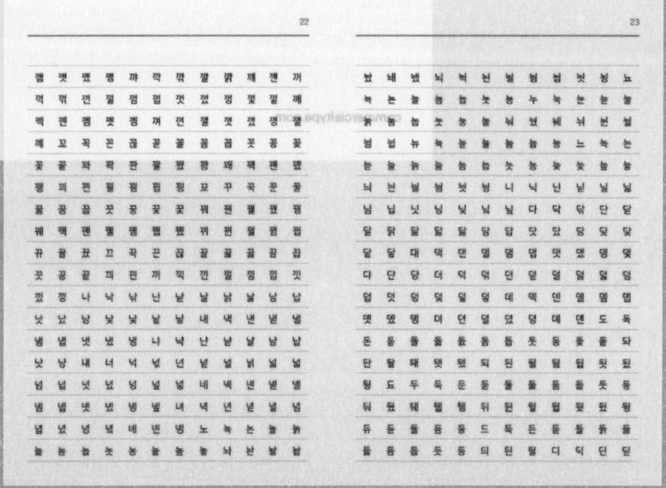

Commercial Type
Typeface: Lyon Arabic Display
Design: Kai Bernau, Waël Morcos, Khajag Apelian
—
commercialtype.com @commercialtype

ليون للعناوين

«ليون عربي» للعناوين هي عائلة خطوط مثالية للعناوين ورؤوس الفقرات. يحافظ «ليون عربي» على تراكيب مُبسَّطة، تمزج بين إنحناءات ناعمة وتفاصيل حادّة الزوايا، مما يعطي الخط شكلاً لينًا وقويًا في آن، يجعله مُريحًا للقراءة، كما هي الحال في نسخة التصميم اللاتينية.

يعتمد التصميم على تباين عالي بين السماكات ومسافات ضيقة بين الحروف وتفاصيل حادّة مما يجعله خيارًا معاصرًا لتطبيقات تيبوغرافيّة شتّى وتبقى مجموعة الأوزان متجذرة في مراجع النسخ التاريخية التي أعيد رسمها لسياق مطبعي معاصر. يتضمن خط «ليون عربي» للعناوين العادي مجموعة كبيرة من الحروف الموصولة التي تحسّن شكل الكلمات وتُسهّل القراءة، كما أنّه يتضمن مجموعة من الأحرف الممدودة المستوحاة من خصائص النسخ والتي تضفي تأثير جمالي على تنسيق الكلمات.

PUBLISHED
2022

DESIGNED BY
KHAJAG APELIAN
WAEL MORCOS
KAI BERNAU

5 STYLES
5 WEIGHTS

FEATURES
PROPORTIONAL OLDSTYLE/LINING FIGURES
FRACTIONS
SUPERSCRIPT/SUBSCRIPT
SWASH ALTERNATES

commercialtype.com

Commercial

6 of 32 — Lyon Arabic Display

تجريد شخصياتهم من إنسانيتهم
برخی گزارش‌ها می‌گوید در یک

الصوت مسموع عن بعد ٤,٦ كيلومتر
حيث يعمد مدرّسو/ات هذه المادة

صرت أقضي وقتًا طويلاً بمفردي
استنكار أساليب اللباس الجديدة

همنوایی شبانه‌ی ارکستر چوب‌ها
وذلك عند ٢٩٨,١٥ كلفن (٢٥°س)

وعُصْبَةٌ يَتَمَيَّزُونَ بِهَا عَنْ غَيْرِهِمْ
اس مرتبہ چار وکٹیں حاصل

commercialtype.com — Commercial

206 pt

Support. Independent. Type II

Contrast Foundry
Typeface: CoFo Peshka
Design: Maria Doreuli
—
contrastfoundry.com @contrastfoundry

FRENCH BY ALAIN DAMASIO

ET SI L'ARÔME DES BOTTES RÉVÈLE MADAME, LE VERSET T'OBSÈDE, MORALISTE!

GERMAN

DIE LIEBE IST SIEGER; STETS REGE IST SIE BEI LEID.

HEBREW

דג סקרן שט בים מאוכזב ולפתע מצא חברה

HUNGARIAN

A LÁNGOLÓ BAZSARÓZSÁRA RÁSZÓR A SZABÓ, LÓG NÁLA.

ICELANDIC

KÆMI NÝ ÖXI HÉR, YKIST ÞJÓFUM NÚ BÆÐI VÍL OG ÁDREPA.

ITALIAN

AI RE SOLO GELATI CON NOCI: TALE GOLOSERIA.

LATVIAN

ČETRI PSIHI FAĶĪRI VĒLU VAKARĀ ZĀĢĒJA GUĻBŪVEI DURVIS, FONĀ ŠŅĀCOT MEŽAM.

LITHUANIAN

ĮLINKDAMA FECHTUOTOJO ŠPAGA SUBLYKČIOJUSI PRAGRĘŽĖ APVALŲ ARBŪZĄ.

MACEDONIAN

ЅИДАРСКИ ПЕЈЗАЖ: ШУГАВ БИЛМЕЗ СО ЧУДЕЊЕ ЏВАКА ЌОФТЕ И КЕЉ НА ТУЃ ЦЕХ.

NORWEGIAN

BORTE PÅ MAGNE-SIA SÅ GREP PER GÅSA I SENGA, MÅPET ROB.

POLISH

A NOEL MUSIAŁ WAPNO ZAWALAĆ? ALA, WAZON PAWŁA I SUM LEONA.

PORTUGUESE

ROMA ME TEM AMOR.

SAMPLES ↓ STYLES: O BLACK, O BLACK SLANTED, O BLACK BACKSLANTED SIZES: 155 PT, 42 PT, 22 PT ↓

FIXED-WING AIRCRAFT
ЛЕОНАРДО ДА ВІНЧІ

THE FRENCHMAN CLEMENT ADER CONSTRUCTED HIS FIRST OF THREE FLYING MACHINES IN 1886, THE ÉOLE. IT WAS A BAT-LIKE DESIGN RUN BY A LIGHTWEIGHT STEAM ENGINE OF HIS OWN INVENTION, WITH FOUR CYLINDERS DEVELOPING 20 HORSEPOWER.

AN AIRCRAFT PROPELLER, OR AIRSCREW, CONVERTS ROTARY MOTION FROM AN ENGINE OR OTHER POWER SOURCE, INTO A SWIRLING SLIPSTREAM WHICH PUSHES THE PROPELLER FORWARDS OR BACKWARDS. IT COMPRISES A ROTATING POWER-DRIVEN HUB, TO WHICH ARE ATTACHED TWO OR MORE RADIAL AIRFOIL-SECTION BLADES SUCH THAT THE WHOLE ASSEMBLY ROTATES ABOUT A LONGITUDINAL AXIS. THREE

САМОЛЕТ Е ВЪЗДУХОПЛАВАТЕЛНО СРЕДСТВО, ПО-ТЕЖКО ОТ ВЪЗДУХА, ОСЪЩЕСТВЯ
ВАЩО ПОЛЕТ В АТМОСФЕРАТА С ПОМОЩТА НА СИЛОВА УСТАНОВКА И НЕПОДВИЖНО
КРИЛО. САМОЛЕТЪТ Е СПОСОБЕН ДА СЕ ПРИДВИЖВА С ВИСОКА СКОРОСТ, ИЗПОЛЗВ
КИ ПОДЕМНАТА СИЛА НА КРИЛОТО ЗА УРАВНОВЕСЯВАНЕ НА ТЕГЛОТО, И ТЕГЛИТЕЛН
СИЛА ОТ СИЛОВАТА УСТАНОВКА ЗА УРАВНОВЕСЯВАНЕ НА ЧЕЛНОТО СЪПРОТИВЛЕНИ

Velvetyne
Typeface: Lithops
Design: Anne-Dauphine Borione
—
velvetyne.fr

Sociotype
Typeface: Onsite Standard
Design: Joe Leadbeater, Nigel Bates
—
socio-type.com @sociotype.foundry

208 pt

Million $$$ Corner

EXPLOITATION

Onsite Standard · Bold · 45pt

'How much space do you need to live?'

Onsite Standard · Bold · 20pt

Oprah struggled to understand how the mother & father and their three daughters managed to cook, eat and sleep - much less have sex, though Oprah was too polite to question - in a room measuring ten by ten feet.

Onsite Standard · Bold · 12pt

For civic authorities, sociologists, urban planners and architects alike, answers the evergreen question of how much domestic space is needed gained even more currency during the COVID-19 lockdowns when entire families were confined for the greater part of the day within the literal four walls of their homes - where they were expected to not just sleep, work, raise and educate children, and socialise, but also to hold onto fraying tempers and unravelling mental issues. The other extreme of the spectrum is those who lived alone, entirely detached from physical touch.

Onsite Standard · Bold · 8?pt · Old Style Numerals

ON A MACRO-LEVEL, they are a timely reminder that the uncontrolled urban sprawl that bedevils almost every single one of the world's 21st-century metropolises need not necessarily be this way'.

These reminders are important because by 2050, seven out of ten people in the world will live in a city, an extraordinary statistic by any metric, never mind the amount of resources that must consume. To literally make room for everyone, the temptation is to just push out beyond existing city boundaries. Never Too Small points out this comes at enormous cost. For far-sighted city architects like ROB ADAMS, who is Melbourne's City Architect, the mission is to build along existing infrastructure and within current boundaries. For every one million people you keep within extant boundaries, you will save approx $nob in new infrastructure costs.

Which means that making do with what we already have - abandoned inner-city warehouses and factories in Boston, and under-utilised shophouses in Singapore, for instance - becomes key. This is certainly not an invitation for developers to continue to churn out expensive but meanly sized cookie-cutter accommodation - the windowless dorms of Munger Hall for UC Santa Barbara students spring to mind here - but rather, it is an opportunity to increase the quality of a development within existing housing stock without sacrificing the humanity of its residents.

All of which brings us back to a full circle to the original question, though lightly amended to account for our mental health and connection to our wider community: "How much space do we really need to live well?" The answer COULD be: "Much less than you think."

capitalist society

Onsite Standard · Thin · 75pt

For a City Dweller

Onsite Standard · Thin · 30pt

In June & July of 2020, there was a 126% increase in people considering homes in village locations.

Onsite Standard · Thin · 45pt

Swapping a zone 2 two-bed for a Cotswold cottage

Onsite Standard · Thin · 20pt

"Pre-pandemic, we were all busy doing a lot of things because it's just what we did, but not necessarily because it's what we wanted to do, or for that matter, what actually made us feel well."
↳ Nancy Colier: Psychology Today

Around Type Family

Around B Black

Around B
Black
Font Size
90pt
Leading
Auto
Tracking
Auto

·Robotic Kitting©

Around B
Black
Font Size
90pt
Leading
Auto
Tracking
Auto

Academic research

Around B
Black
Font Size
90pt
Leading
Auto
Tracking
Auto

Häfen und Bahnhöfe

www.estudiocalderon.com 9 of 17

Around A Type Family calderon

• The first parameter is dedicated to the connection of Arabic culture with the world by subtracting the obvious and adding the specific.

• The second parameter seeks to achieve simplicity through reduction that appeals to a friendlier and more contemporary aspect.

• The third parameter explores the rythym of counterforms of the letters of the road signs of "West Berlin".

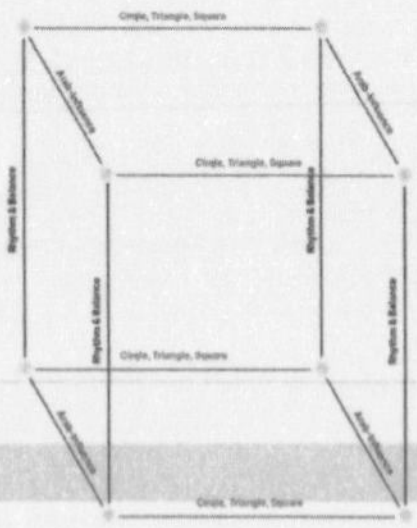

www.estudiocalderon.com 4 of 34

Estudio Calderón
Typeface: Around A & Around B
Design: Felipe C. Arteaga
—
estudiocalderon.com @calderon_type

Around Type Family calderon

Around B: Family Overview

Around B Light
ABCDEFGHIJKLMNOPQRSTUVWXYZ
abcdefghijklmnopqrstuvwxyz ŀiŀtŀti
0123456789 & .,:;!?(),„"«»+@©®™

Around B Regular
ABCDEFGHIJKLMNOPQRSTUVWXYZ
abcdefghijklmnopqrstuvwxyz ŀiŀtŀti
0123456789 & .,:;!?(),„"«»+@©®™

Around B Medium
ABCDEFGHIJKLMNOPQRSTUVWXYZ
abcdefghijklmnopqrstuvwxyz ŀiŀtŀti
0123456789 & .,:;!?(),„"«»+@©®™

Around B Bold
ABCDEFGHIJKLMNOPQRSTUVWXYZ
abcdefghijklmnopqrstuvwxyz ŀiŀtŀti
0123456789 & .,:;!?(),„"«»+@©®™

Around B Black
ABCDEFGHIJKLMNOPQRSTUVWXYZ
abcdefghijklmnopqrstuvwxyz ŀiŀtŀti
0123456789 & .,:;!?(),„"«»+@©®™

www.estudiocalderon.com 4 of 17

Around A Type Family calderon

Around A Medium

Johanna Moosdorf

STARSHIP UPDATE.

Philipp Holfmann

www.estudiocalderon.com 12 of 34

WEIGHT
Regular

SIZE
170 pt

LEADING
172 pt

TRACKING
Default

LANGUAGE
English

OPENTYPE
Turn off

Trilling Gallant *MODI.

Estudio Calderón
Typeface: Chroma Sans
Design: Felipe C. Arteaga
—
estudiocalderon.com @calderon_type

Fatype
Typeface: Make Way
Design: Yassin Baggar
—
fatype.com @fatypefonts

Make Way
Black

Make Way Fatype

ARMADA Secretive

94 pt

VAGABOND Point Blanks Mechanized

72 pt

CONCEPTUALS Echelön; (128)GB

54 pt

NEW HELIOCENTRIC London Protect Sacral

40 pt

DRAMAS & VINTAGE WINERY €0.783 Lemonade Stand Bully,

30 pt

SUPERCONTINENTE
Typeface: Montiac
Design: Fabiola Mejía
supercontinente.com @supercontinente_

MONTIAC REGULAR | ENGLISH | 68PT

What is a horn? It is a *mountain peak* formed by headward erosion of multiple glaciers, identified in the 1920s.

MONTIAC REGULAR | ENGLISH | 120PT

Tel. +1-865-999-532

MONTIAC ITALIC SEMI MONO | LATVIAN | 48PT

Šeit uzskatāmi tiek atainota sarežģītā un miljardiem gadu ilgā Latvijas teritorijas veidošanās. Informāciju papildina ģeoloģiskā karte, kas atspoguļo nogulumiežu izplatību

MONTIAC SEMI MONO | SPANISH | 100PT

₡250: Precio de feria

SUPERCONTINENTE.COM MONTIAC TYPE SPECIMEN SAN JOSÉ, COSTA RICA

MONTIAC | SPECIFIC... →CHARACTER SET
OPEN TYPE ... GE SUPPORT

UPPERCASE
AÁĂÂÄÀĀĄÅÃÆBC
FGĞĢĠHĦIÍÎÏÌĪĮJ
ÖÒŐŌØÕŒPÞQRŘ
ŨŬŲŮVWẂŴẄẀX

LOWERCASE
aáăâäàāąåãæbcćç
hħiíîïìīįjjkķlĺľļłmnń
qrŕřŗsśšşßtťţţuúûüù
ÿỳzźžż

CURRENCY
₿¢₡¤$€ƒ₺£₩¥

LINING FIGURES
01234567890

OLDSTYLE FIGURES
01234...

PUNCTUATION
.,:;!¿?·•*#/\(){}[]

CASE SENSITIVE FORMS
·•/\(){}[]-–—«»

SYMBOLS AND ORDINALS
%‰@&¶§©™°+‡

DIACRITICAL SIGNS

FRACTIONS
/ ½ ⅓ ⅔ ¼ ¾ ⅕ ⅖ ⅗ ⅘ ⅙ ⅚

NOMINATOR AND DENOMINATOR | SUBSCRIPT AND SUPERSCRIPT
0123456789H0123456789 0123456789H0123456789

ARROWS
↑↗→↘↓↙←↖↔↕

SC-T01

ABOUT MONTIAC

A RATIONAL SERIF AT HEART, MONTIAC COMPRISES A PROPORTIONAL CUT, AN ACCOMPANYING ITALIC, AND A MONO. SPECIFICALLY ROOTED IN THE ANGULARITY OF COLUMBIA ANTIQUA (1893), THE FAMILY REINTERPRETS ITS FEATURES AND DISCOVERS TROPICAL ALPINISM.

STYLES: REG SEMI-MONO MONO ITALIC

SC-T01

RAW/WORLD DATA:

RELEASE DATE: 2021

DETAILS:

SC-T01

..., Catalan, Cebuano, Chiga, Colognian, Cor... Estonian, Faroese, Filipino, Finnish, French, ...ndic, Ido, Inari Sami, Indonesian, Interlin... Kalenjin, Kinyarwanda, Kurdish, Latvian, ...embourgish, Luyia, Machame, Makhu... Morisyen, North Ndebele, Northern ...orsk, Nyanja, Nyankole, Occitan, Oromo, ...a, Samburu, Sango, Sangu, Sardinian, ..., Soga, Somali, South Ndebele, Southern ...aita, Taroko, Teso, Tsonga, Tswana, Turk... ...tern Frisian, Wolof, Xhosa, Zulu

...by the terms of the ...CONTINENTE (Fabiola Mejía).

Special thanks to Federico Parra for his production assistance and Ryan Bugden for his feedback in the process.

 CONTACT FABI@SUPERCONTINENTE.COM FOR ANY QUESTIONS SAN JOSÉ, COSTA RICA

فانيلا (نبات)

نبتة دائمة الخضرة.
الثمرة عبارة عن قرن.

2022-02-25 - 6

Heheh-Gawhar-v02_Specimen-Print

اللون الأحمر أحد الألوان الأولية الثلاثة للضوء المرئي، (وهي: الأحمر، والأخضر، والأزرق؛ وتختصر بالإنجليزية) وأيضا في الألوان الأولية المادية (وهي: الأحمر، والأصفر، والأزرق). ويتراوح طول موجة اللون الأحمر بين ٦٢٥—٧٦٠ نانومتر وتعرف أطوال الموجات الأقل من ذلك

ءأإآا تتثيت ت جحج جذذ زز
ششش ضضض ظط غغغ غ
ففف ف ككك ك االل ل ق ق
ممم م ن ن ووؤ ههه ه يي ي لا
لا : ؛ ، / « ؟ ! » ١ ٢ ٣ ٤ ٥ ٦ ٧ ٨

الأرجواني

120pt
بونت ١٢٠

صبغات عاكسة
الألوان الساخنة

70/70pt
بونت ٧٠/٧٠

عند إنتاج لون دهان أو دهان
سطح ما، فإن اللون يغير شكل

38/40pt
بونت ٣٨/٤٠

**لون الطاقة والحماس، والسعادة، والطاقة،
والتوازن، والحرارة، والحرائق، والحماس،
التوهج، واللعب، البهرجة، الخريف، والرغبة**

26/30pt
بونت ٢٦/٣٠

2022-02-25 - 2

Heheh-Gawhar-v02_Specimen-Print

Heheh Type
Typeface: Gawhar
Design: Heheh Type
—
hehehtype.com @hehehtype

S.M. Foundry
Typeface: Maxeville
Design: Mark Niemeijer
—
s-m.nu @sm.foundry

Proportional, 10 Styles.

1 Maxeville Tekst + 2 *Italic*
3 Maxeville Regular + 4 *Italic*
5 Maxeville Construct + 6 *Italic*
7 **Maxeville Bold** + 8 ***Italic***
9 **Maxeville Bold Construct** + 10 ***Italic***

Monospace, 4 Styles.

11 Maxeville Regular
12 Maxeville Construct
13 **Maxeville Bold**
14 **Maxeville Bold Construct**

14 pt International Monument Foundation Structuralist Bureau Retrospective
24 pt Olympic Metabolism Expression After Jazz
34 pt Rational Thoughts Utilize Formats
44 pt Offices For The Generation
54 pt Wealth Reflects Phrases
64 pt Futurist Institute Tasks
74 pt Artificial Manifesting
84 pt Globalize Objective
94 pt Villager Associates

216 pt

Out of the Dark
Typeface: Cosplay
Design: Philipp Herrmann
outofthedark.swiss @outofthedark.xyz

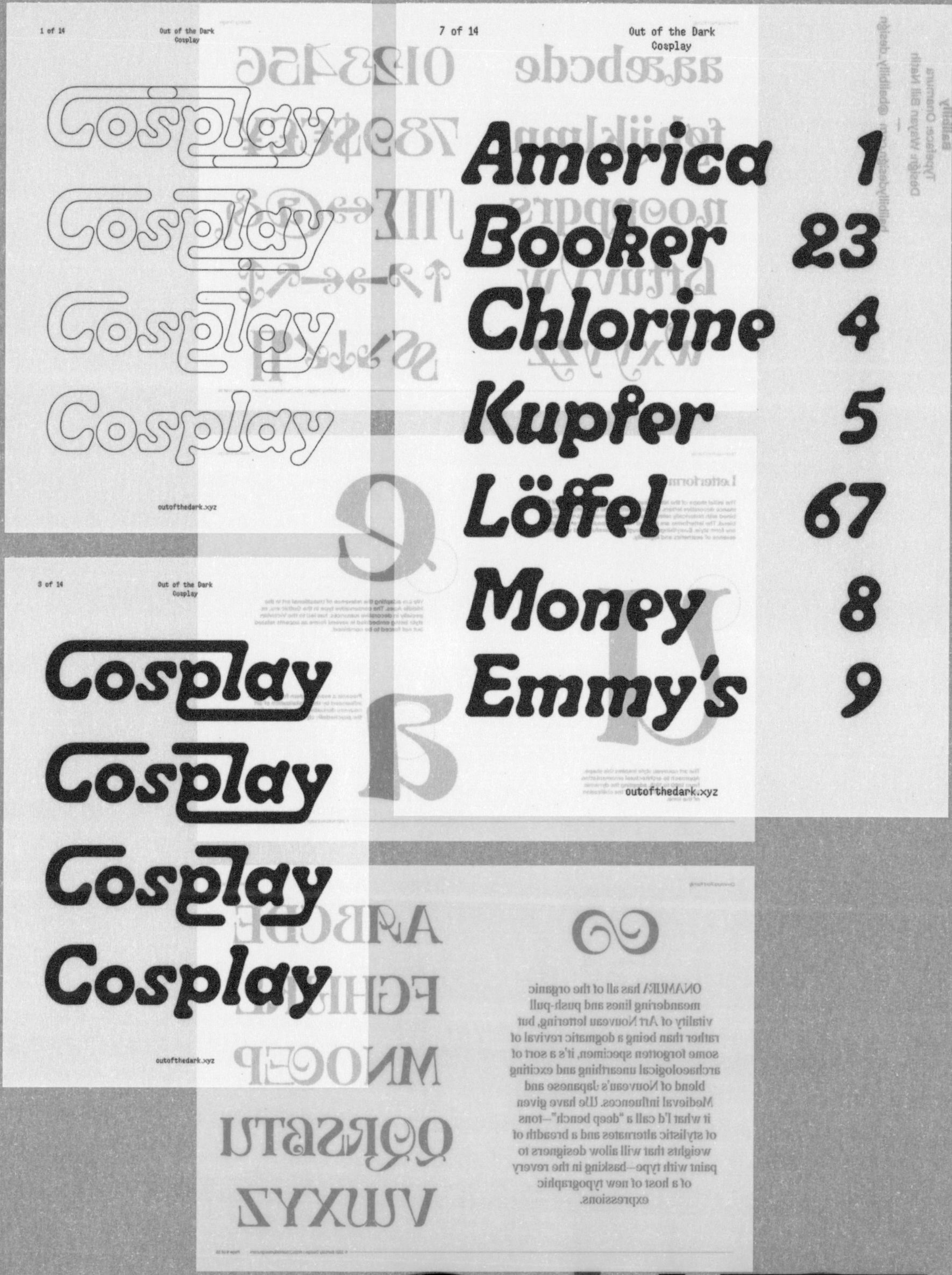

Balibilly
Typeface: Onamura
Design: Wayan Bill Natih
—
balibillydesign.com @balibilly_design

SUVA Type Foundry
Typeface: Esimene, Bomber Fraktur, Võrsik
Design: Andree Paat, Mathias Väärsi, Karri Kaljend
—
balibillydesign.com @balibilly_design

Support Independent Type II

Esimene regulaarne 2014 Andree Paat

ABCDEFGHIJKLMNO

0123456789

Emoji are ideograms and smileys used in electronic messages and web pages.

PQRSZTUVWÕÄÖÜXY

2

Quirky geometric typeface based on 5 random objects.

Bomber Fraktur 2015 Mathias Väärsi

46

If you were a steak you would be well done.

Võrsik 2015 Karri Kaljend

ABCDEFGHIJKLMNO

0123456789

Kurita took inspiration from weather forecasts that used

PQRSZTUVWÕÄÖÜXY

14

Typeface that tries to capture movement and speed.

Affichen-Schriften, 1821–1850 forgotten-shapes.com

Schmale Egyptienne No. 12, 562 pt

Fabrik

Affichen-Schriften, 1821–1850 forgotten-shapes.com

Schmale Egyptienne No. 12, 130 pt

HÅNDVÆRK
Přibližně

Schmale Egyptienne No. 12, 78 pt

LÉON PIERRE-QUINT
Skjöl orðið skylt
Johann Heinrich Merck

Schmale Egyptienne No. 12, 78 pt

CROISIÈRE D'HIVER
Bd. 1. Berlin 1835
»Neues Wiener Journal«

Schmale Egyptienne No. 12, Stylistic Set 03, 130 pt

AURĄ DZIEŁA
Jeanne d'Arc

Forgotten Shapes
Typeface: Schmale Egyptienne No. 12
Design: E. Tarbé & Cie. (Fonderie Génerale)
—
forgotten-shapes.com @forgotten_shapes

Inari Type
Typeface: Nikkei Maru
Design: Caio Kondo, Francesca Bolognini, Mat Desjardins
—
balibillydesign.com @balibilly_design

Slow Fonts
Typeface: Method Mono
Design: Fabien Coupas
—
slowfonts.xyz @slowfonts

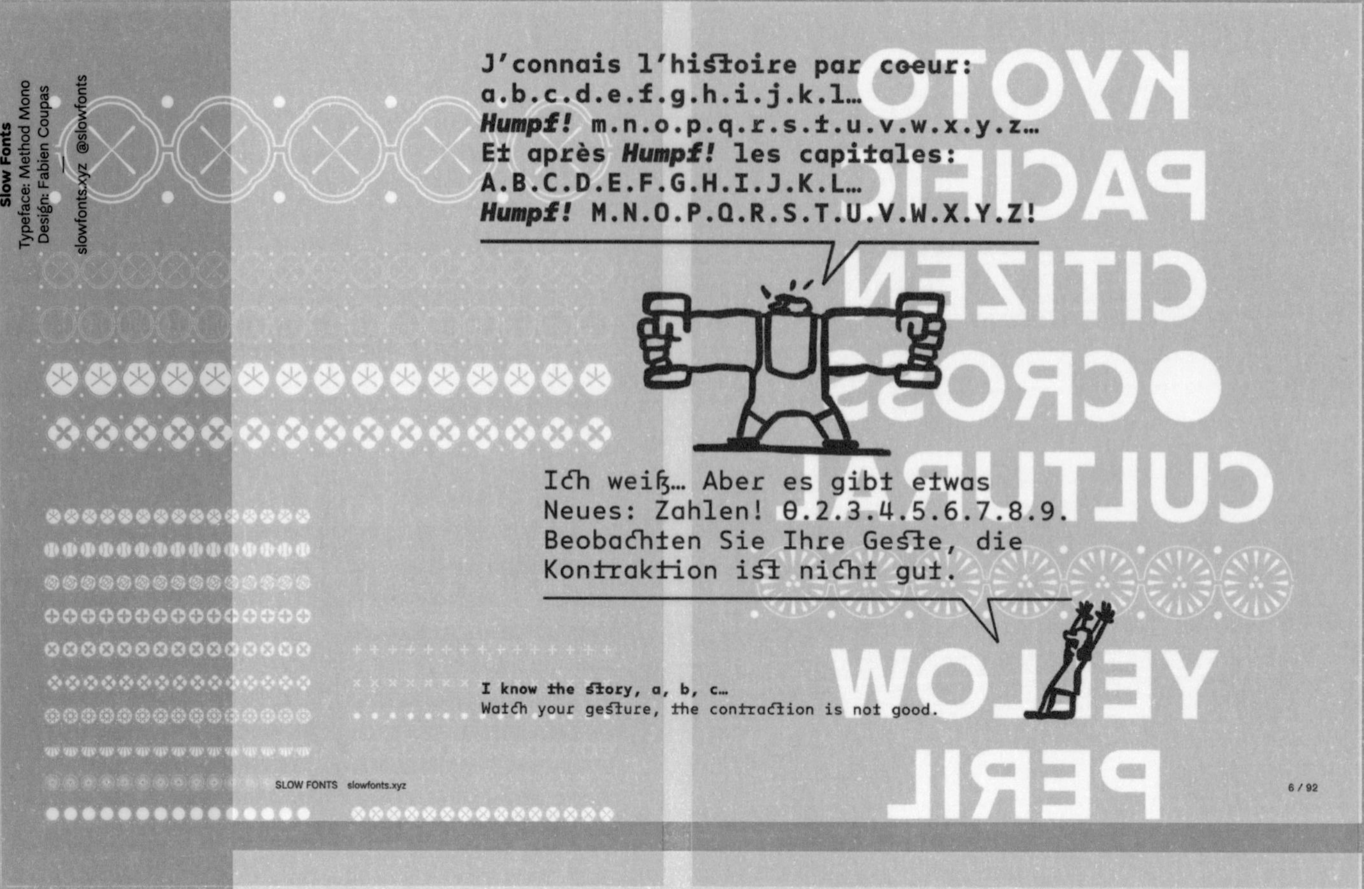

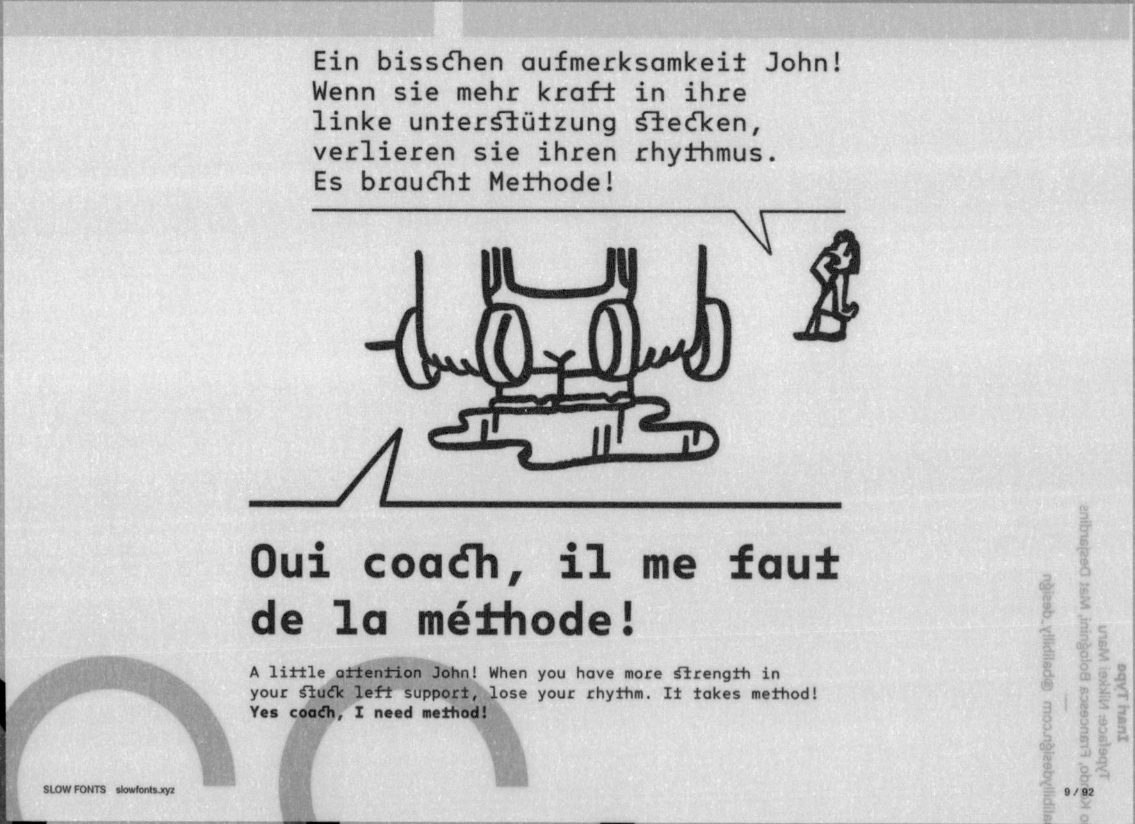

222 pt

Support: Independent Type II

figure
étoffe
Pour apprendre à lire,
il faut d'abord lire très lentement.
Il ne faut pas avoir de paresse
en lisant. Ni de précipitation.
La précipitation n'est d'ailleurs
qu'une autre forme de la paresse.
Gymnaste
triomphe
Sifflets
Grille
Motivé
Auguste Émile Faguet,
Écrivain & critique
littéraire français.
1847 - 1916

224 pt

Wise Type
Typeface: WT Zaft
Design: Jacob Wise
—
wisetype.nl @wise_type

The Temporary State
Typeface: Kommuna
Design: Roman Gornitsky
—
type.tmpstate.net

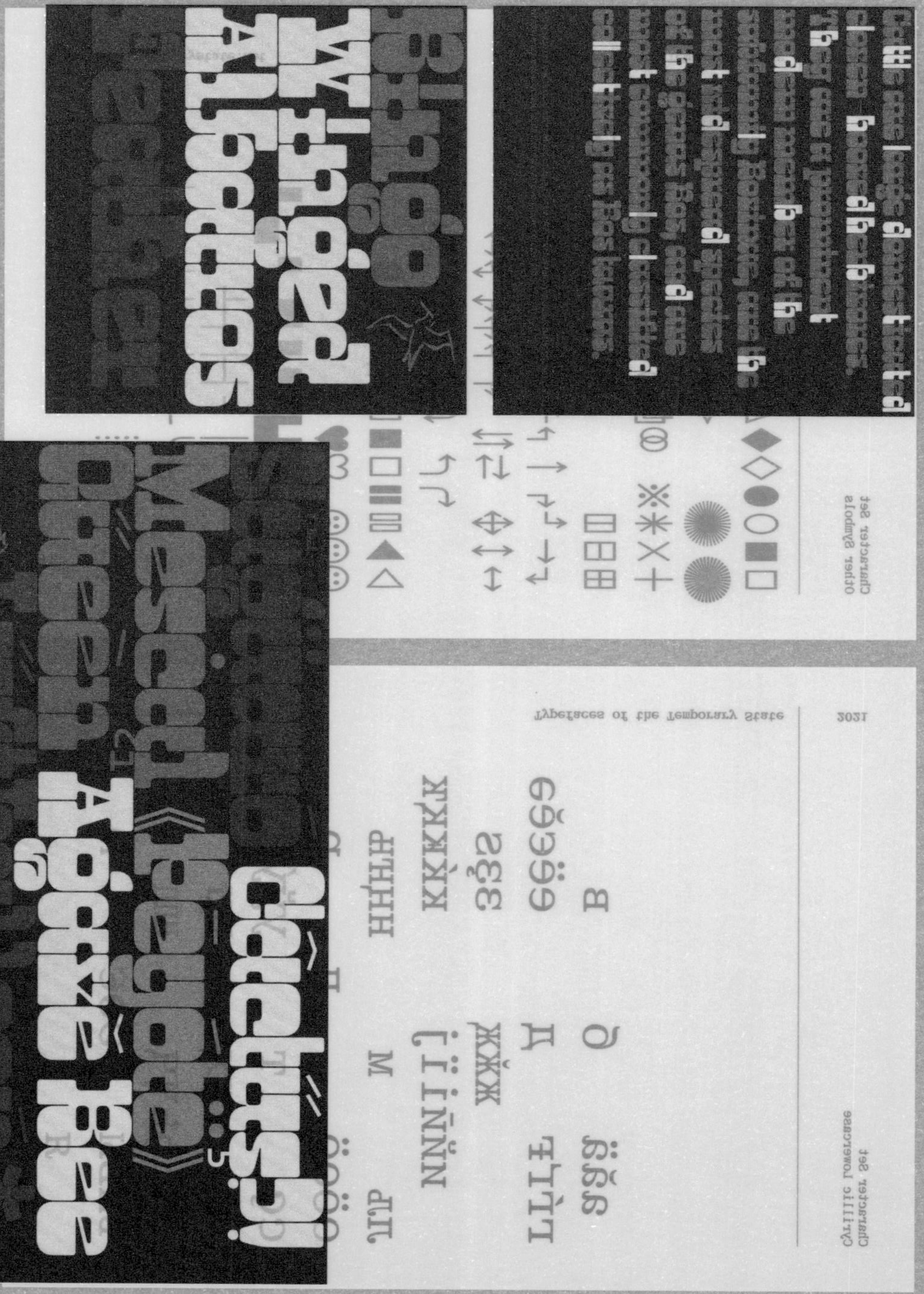

The Temporary State
Typeface: Kommuna
Design: Roman Gornitsky
—
type.tmpstate.net

Mass-Driver
Typeface: MD Polychrome
Design: Rutherford Craze
mass-driver.com @mass.driver

MD Polychrome — Styles

Nano — 160PT
Venus 1986

Nano — 100PT
201 GHz Resolution

Nano — 60PT
M74 [Phantom Galaxy], Group 3

Nano — 40PT
Saturn (9.5 AU {1.426 km; 880m mi} from Sol)

MD Polychrome © Mass-Driver 2022 — mass-driver.com — 5 / 16

KONVOLUTE Entanglement
INSCRIPTION Gaďadykkus
DRUKKERIJ Atmospheres
HYPERCUBES Reingestoßen
MICROTESLA Brainstorming
GEZWITSCHER Zwyciężyłem

Now Showing:
Stable Diffusion AI "Art"
32 (The Nevada Monolith) – Unknown

A51 North →
Newfoundland & Lb.
86% of $21.99, and Other Queries
Magnetic Ink Character Recognition (MICR) uses a

HARDWARE Megawatts
DISCOVERY Gekopiëerd
ARMOURED Avalanches
BØRSKRAK Hafrajmjölið
LIQUIDITÉS Retroactive
TAXONOMY Kalizjańską

OLOF
Typeface: Linea 72
Design: Roland Hirter, 1972 / Thomas Hirter, 2022
olof.ch @thomas_hirter

Linea 72
Technical Information & Type Specimen

Linea 72 Kontur

210 pt

Zirkel

124 pt (ss03)

GEOMETRY

132 pt (ss03, ss04)

Reissfeder

92 pt

drawing tools

56 pt

pencil and ink on paper

Studio Feixen
Typeface: Studio Feixen Family
Design: Felix Pfäffli, Robin Eberwein
fonts.studiofeixen.ch @studiofeixen

228 pt

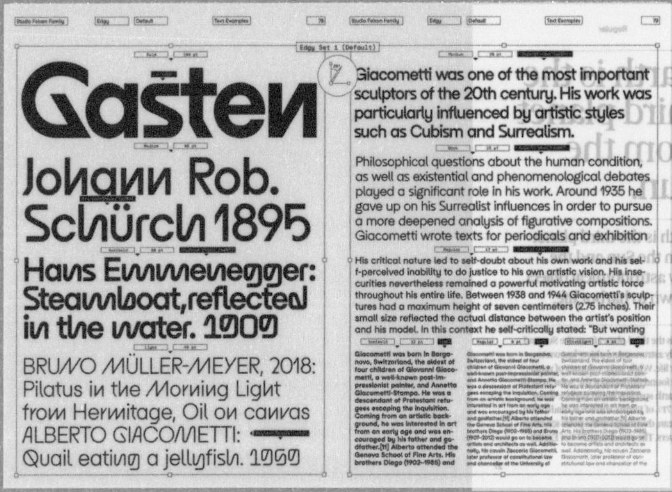
Gasten
Johann Rob. Schürch 1895
Hans Emmenegger: Steamboat, reflected in the water. 1909
BRUNO MÜLLER-MEYER, 2018: Pilatus in the Morning Light from Hermitage, Oil on canvas
ALBERTO GIACOMETTI: Quail eating a jellyfish. 1959
Giacometti was one of the most important sculptors of the 20th century. His work was particularly influenced by artistic styles such as Cubism and Surrealism.
Philosophical questions about the human condition, as well as existential and phenomenological debates played a significant role in his work. Around 1935 he gave up on his Surrealist influences in order to pursue a more deepened analysis of figurative compositions. Giacometti wrote texts for periodicals and exhibition
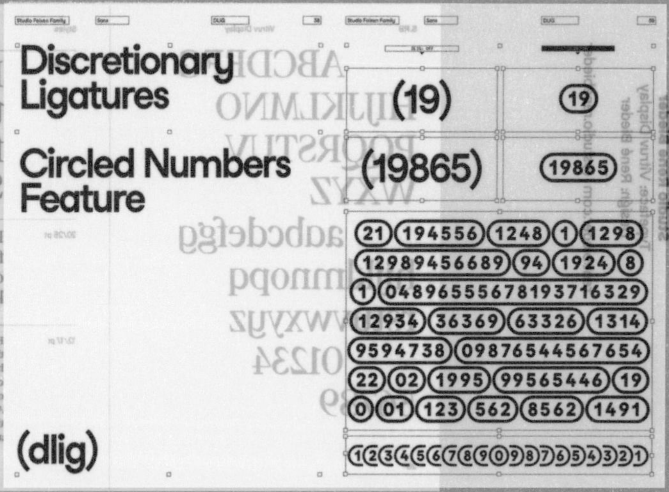
Discretionary Ligatures
Circled Numbers Feature
(dlig)
(19)
(19865)
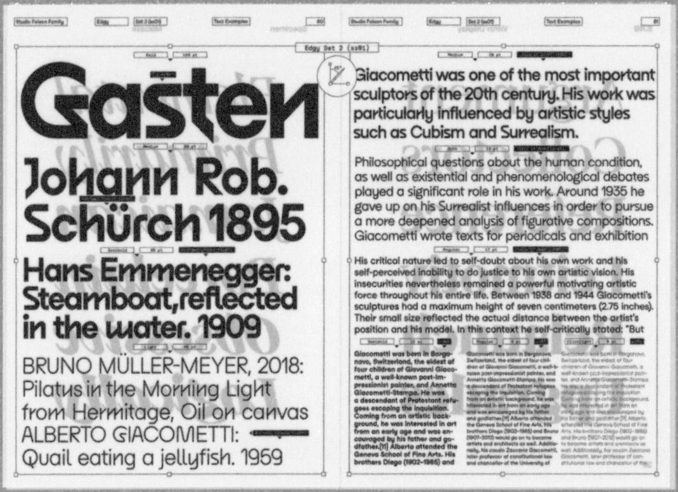
Gasten
Johann Rob. Schürch 1895
Hans Emmenegger: Steamboat, reflected in the water. 1909
BRUNO MÜLLER-MEYER, 2018: Pilatus in the Morning Light from Hermitage, Oil on canvas
ALBERTO GIACOMETTI: Quail eating a jellyfish. 1959
Giacometti was one of the most important sculptors of the 20th century. His work was particularly influenced by artistic styles such as Cubism and Surrealism.
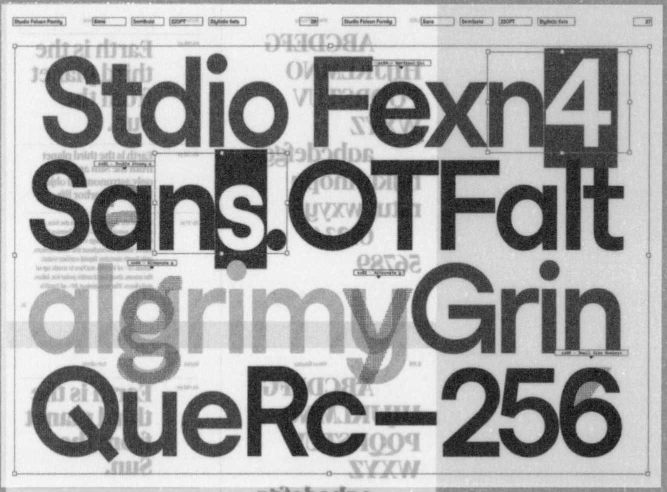
Stdio Fexn4
Sans.OTFalt
algrimyGrin
QueRc–256
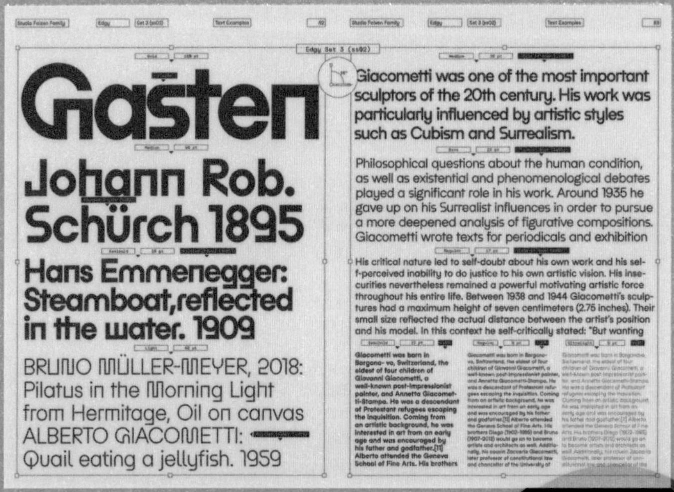
Gasten
Johann Rob. Schürch 1895
Hans Emmenegger: Steamboat, reflected in the water. 1909
BRUNO MÜLLER-MEYER, 2018: Pilatus in the Morning Light from Hermitage, Oil on canvas
ALBERTO GIACOMETTI: Quail eating a jellyfish. 1959
Giacometti was one of the most important sculptors of the 20th century. His work was particularly influenced by artistic styles such as Cubism and Surrealism.

01234
56789

Studio Rene Bieder
Typeface: Vitruv Display
Design: René Bieder
renebieder.com @studio.renebieder

230 pt

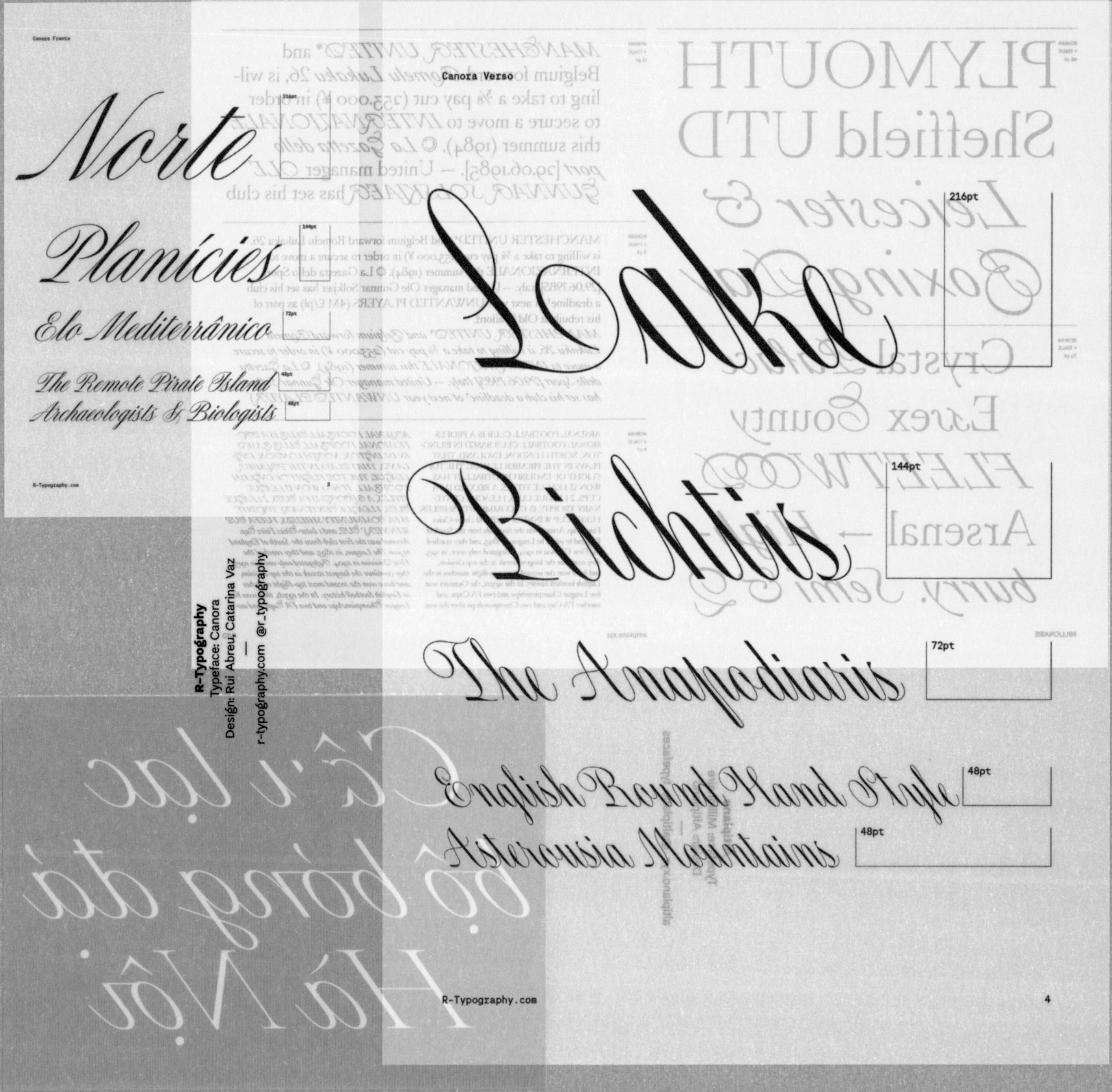
Canora Frente
Canora Verso
Norte
Planícies
Elo Mediterrânico
The Remote Pirate Island
Archaeologists & Biologists
Lake
Richtis
The Anapodiaris
English Round Hand Style
Asterousia Mountains
PLYMOUTH
Sheffield UTD
Leicester &
Arsenal
216pt
144pt
72pt
48pt
R-Typography.com
4
R-Typography
Typeface: Canora
Design: Rui Abreu, Catarina Vaz
—
r-typography.com @r_typography

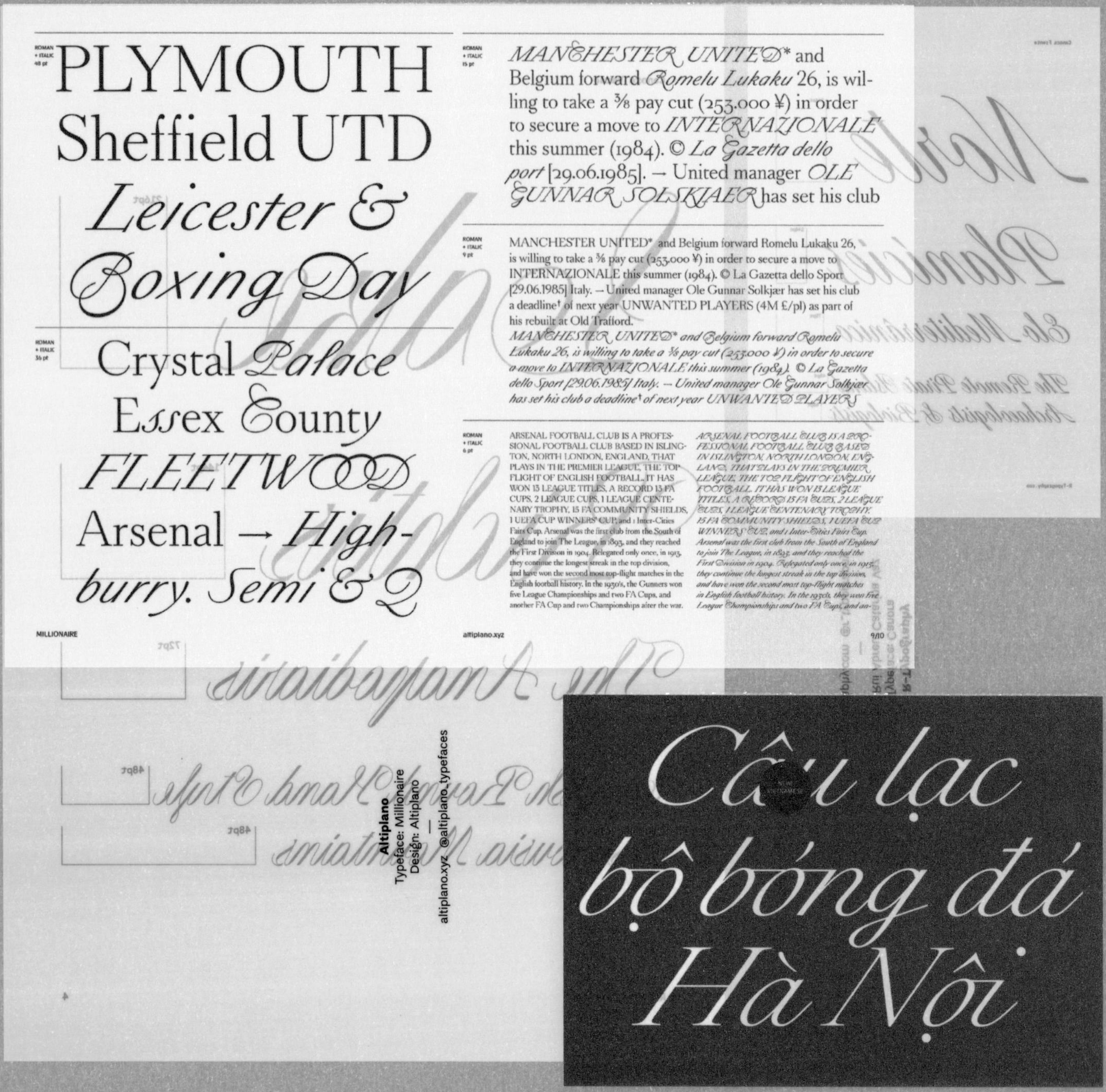

Altiplano
Typeface: Millionaire
Design: Altiplano
—
altiplano.xyz @altiplano_typefaces

232 pt

Bastarda Type Foundry
Typeface: Salsa BT
Design: Jason Guzmán, Seb Castellanos de la Hoz
—
bastardatype.com @bastardatype

Bastarda Type Foundry®

Salsa BT
Typeface Quick Manual

Use of Font Overview

1981

Salsa BT Round, 177 pt

OBLIGADO A CANTAR
TODA LA NOCHE

Salsa BT Round, 35 pt

NEW YORK

Salsa BT Regular, 73 pt

COLÒN

Salsa BT Round, 108 pt

Li

Salsa BT Ligature, 328 pt

HECTOR

Salsa BT Round, 35 pt

IDILIO

Salsa BT Round, 342 pt

bastardatype.com

What *was* I *after* all?

“Yesterday the poet said life is a vale of tears; *today* he said life is a land of smiles; and he was right both times.”

Game

Cyrillic Overview
Regular • Bold • Thin • Extrabold
Люблю
Россия омывается тремя океанами: Атлантическим, Тихим и Северным Ледовитым.
На территории России расположены несколько горных цепей3: Урал, Кавказ, Алтай и другие.
Ж
Margit
2021
Schriftlabor
Typeface: Margit
Design: Miriam Surányi
—
schriftlabor.at @schriftlabor
WHOA TOP
ULTRASONIC
REFRACTION
COMPRESSION
OSCILLATION
FREQUENCY
WAVELENGTH
Cyrillic (Bulgarian Localization) Overview
Light • Extrabold • Semibold • Thin
Началното образование
Всеки човек има право на образование.
Член
Образованието трябва да бъде безплатно, поне що се отнася до началното и основното
Margit
2021
Overview
Extrabold • Regular • Bold • Medium
Quest*
He looked long and hard in his youth, but that role model never materialized. His only choice was to embrace all the people in his life he didn't want to be like.
"The question is in a way meaningless, she knows, but one must ask. Love in such situations is rarely real. Sex is the engine, exalting and ruining people, sex and frustration. Love is what people believe is worth the path of devastation."
189
Margit
2021

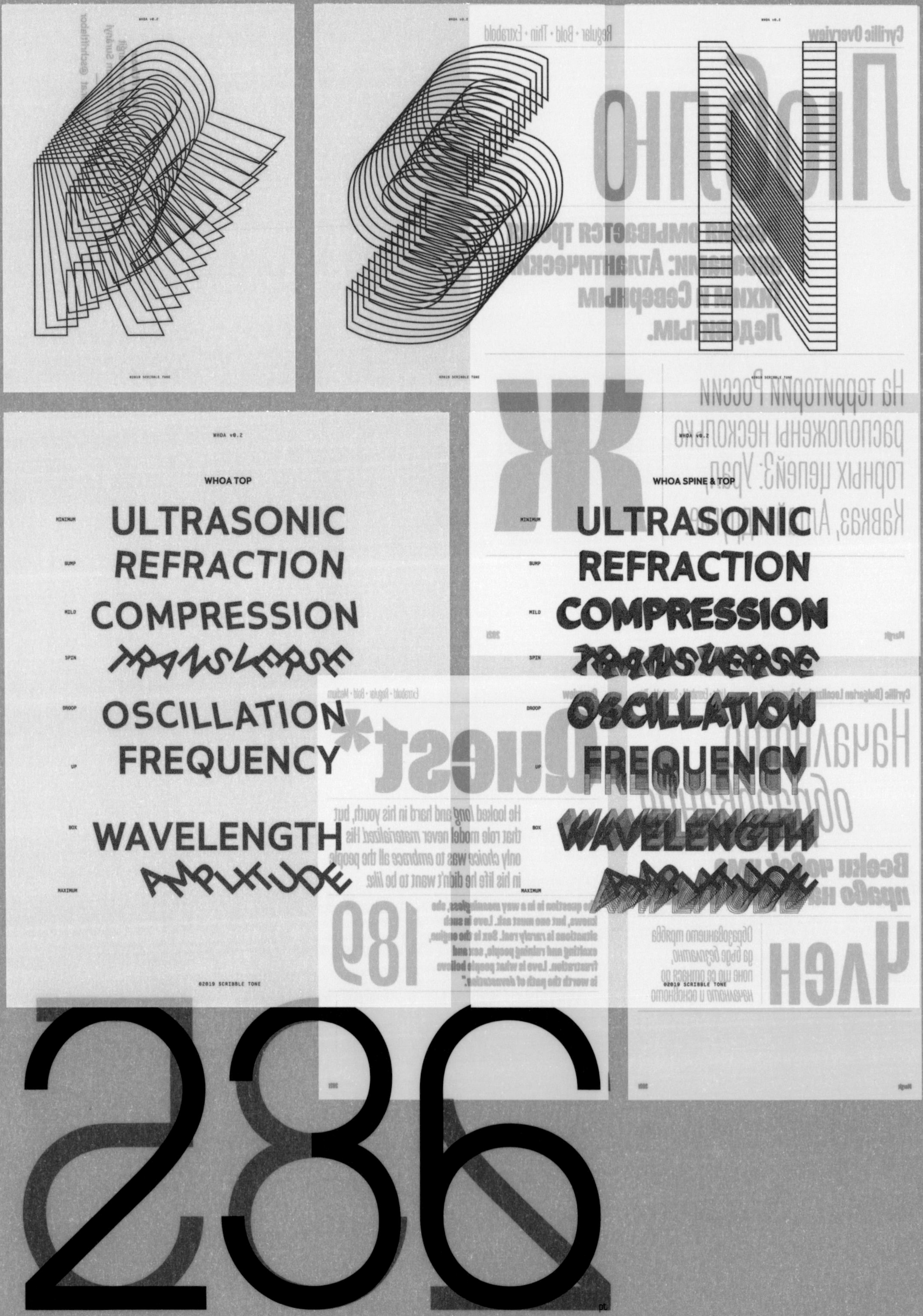
WHOA v0.2
WHOA TOP
MINIMUM
ULTRASONIC
BUMP
REFRACTION
MILD
COMPRESSION
SPIN
TRANSVERSE
DROOP
OSCILLATION
UP
FREQUENCY
BOX
WAVELENGTH
MAXIMUM
AMPLITUDE
©2019 SCRIBBLE TONE
WHOA SPINE & TOP
286
pt

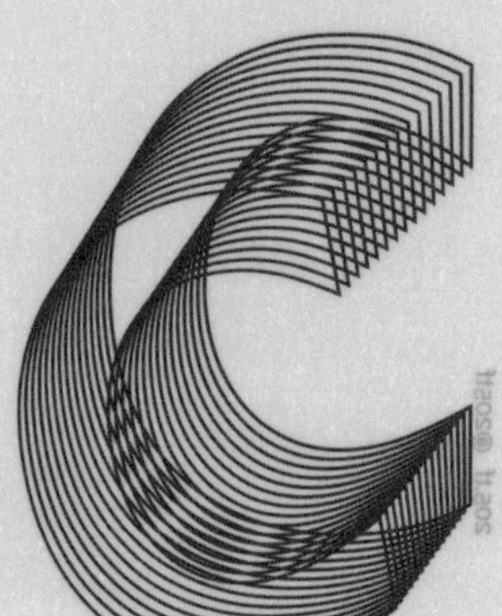

Scribble Tone
Typeface: WHOA
Design: Scribble Tone
—
scribbletone.com @scribbletone

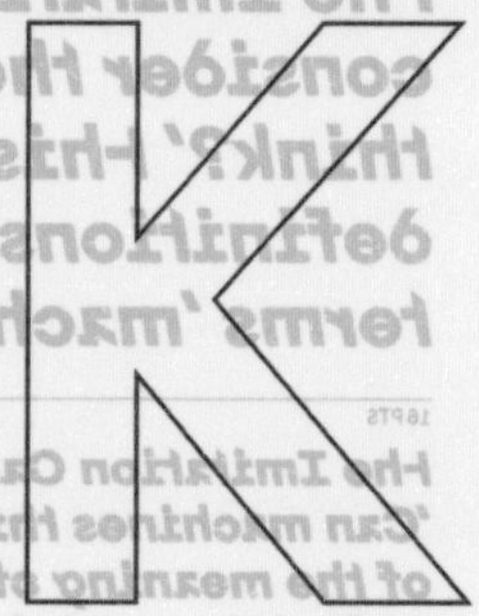

AUGURE SIMON RENAUD 2023

BOLD SLANTED

56 PTS

The Imitation Game. I propose to consider the

32 PTS

The Imitation Game. I propose to consider the question, 'Can machines think?' This should begin

24 PTS

The Imitation Game. I propose to consider the question, 'Can machines think?' This should begin with definitions of the meaning of the terms 'machine' and 'think'. The

16 PTS

The Imitation Game. I propose to consider the question, 'Can machines think?' This should begin with definitions of the meaning of the terms 'machine' and 'think'. The definitions might be framed so as to reflect so far as possible the normal use of the words, but this attitude is dangerous. If the meaning of the words 'machine' and 'think' are to be found by examining how they are commonly used it is difficult to escape the conclusion that the meaning and the answer to the question, 'Can

205TF © 2023-02 99/123

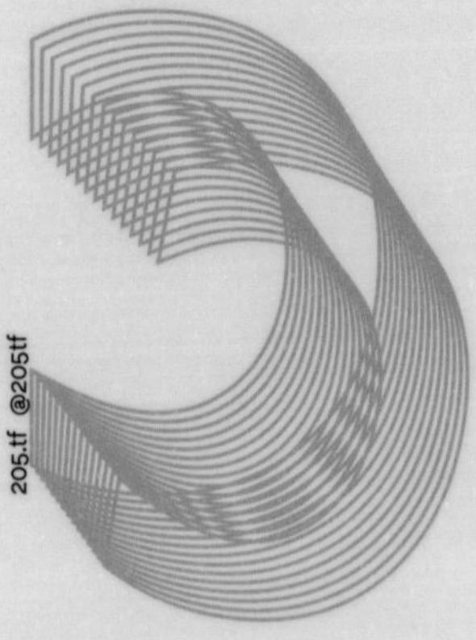

205TF
Typeface: Augure
Design: Simon Renaud
—
205.tf @205tf

AUGURE SIMON RENAUD 2023

MEDIUM

56PTS

The Imitation Game. I propose to consider the

32PTS

The Imitation Game. I propose to consider the question, 'Can machines think?' This should begin

24PTS

The Imitation Game. I propose to consider the question, 'Can machines think?' This should begin with definitions of the meaning of the terms 'machine' and 'think'. The

16PTS

The Imitation Game. I propose to consider the question, 'Can machines think?' This should begin with definitions of the meaning of the terms 'machine' and 'think'. The definitions might be framed so as to reflect so far as possible the normal use of the words, but this attitude is dangerous. If the meaning of the words 'machine' and 'think' are to be found by examining how they are commonly used it is difficult to escape the conclusion that the meaning and the answer to the question, 'Can

205TF © 2023-02 75/123

AUGURE SIMON RENAUD 2023

EXTRALIGHT SLANTED

56PTS

The Imitation Game. I propose to consider the

32PTS

The Imitation Game. I propose to consider the question, 'Can machines think?' This should begin with definitions of the

24PTS

The Imitation Game. I propose to consider the question, 'Can machines think?' This should begin with definitions of the meaning of the terms 'machine' and 'think'. The definitions

16PTS

The Imitation Game. I propose to consider the question, 'Can machines think?' This should begin with definitions of the meaning of the terms 'machine' and 'think'. The definitions might be framed so as to reflect so far as possible the normal use of the words, but this attitude is dangerous. If the meaning of the words 'machine' and 'think' are to be found by examining how they are commonly used it is difficult to escape the conclusion that the meaning and the answer to the question, 'Can machines think?' is to be sought in a

205TF © 2023-02 37/123

238

pt

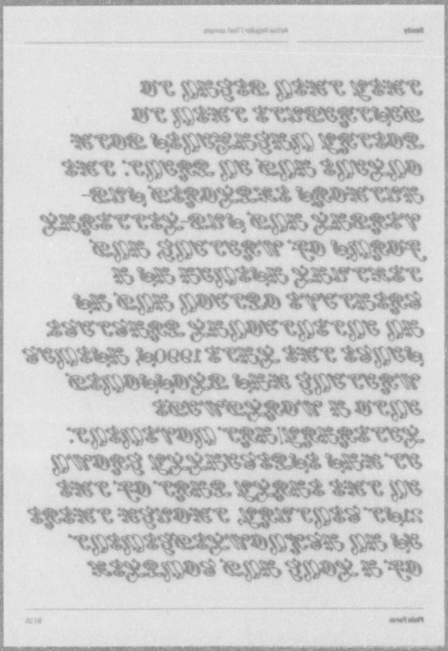

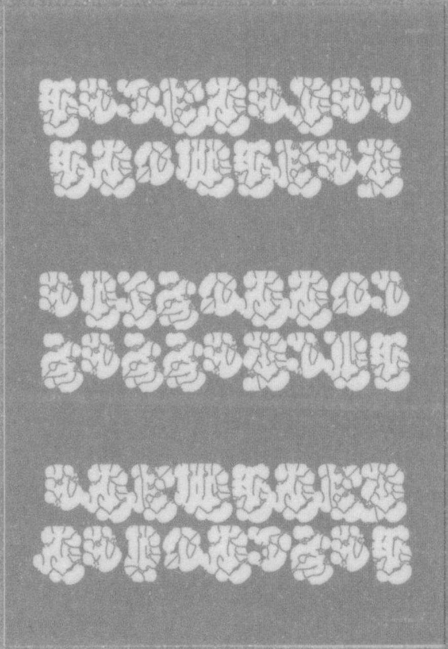

AUGURE
SIMON RENAUD
2023
AUGURE TYPEFACE
240PTS
Aug
120PTS
Augure
56PTS
Augure Augure A
32PTS
Augure Augure Augure Augure
24PTS
Augure Augure Augure Augure Augure A
16PTS
Augure Augure Augure Augure Augure Augure Augure Augur
205TF
© 2023-02
2/123

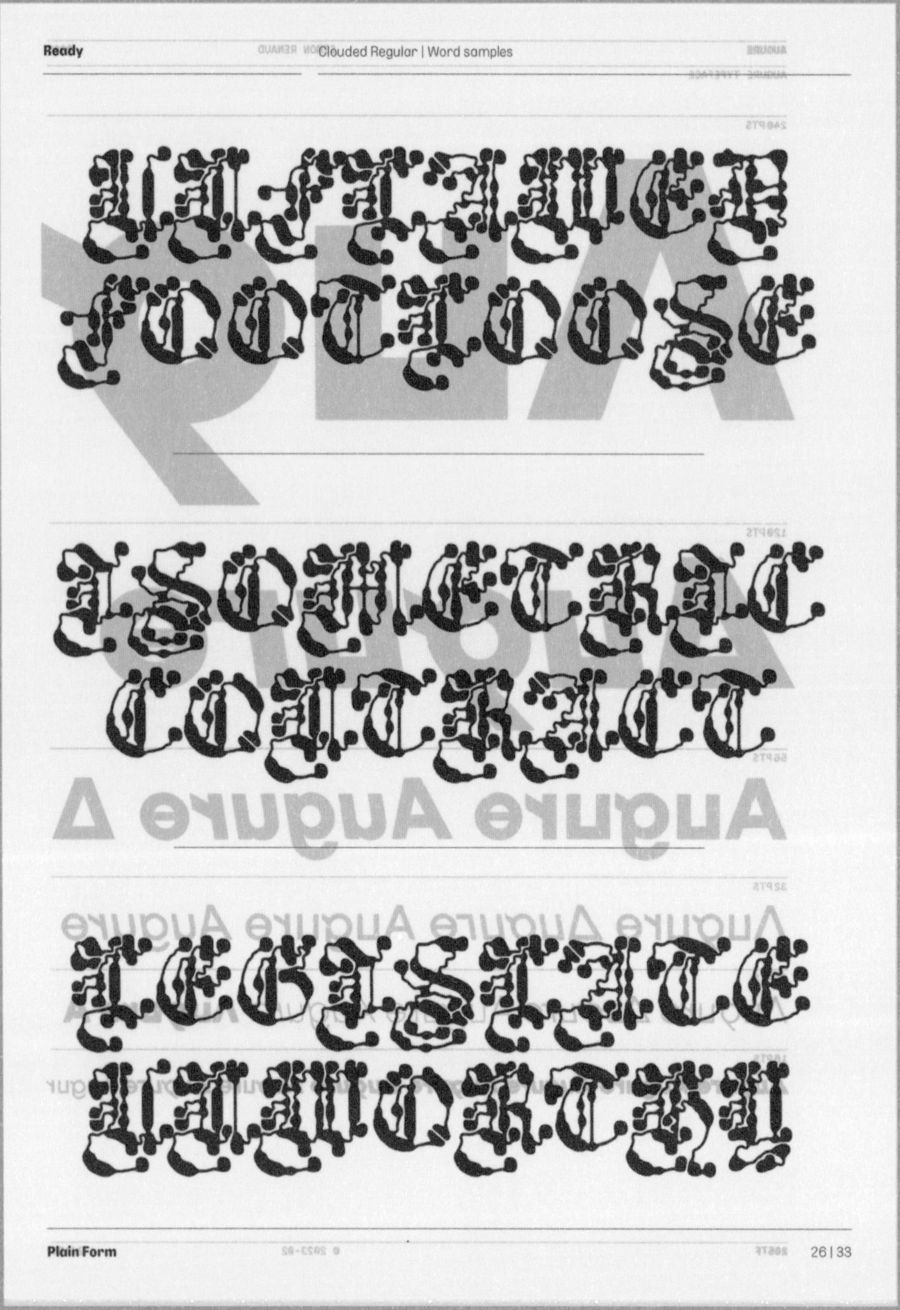

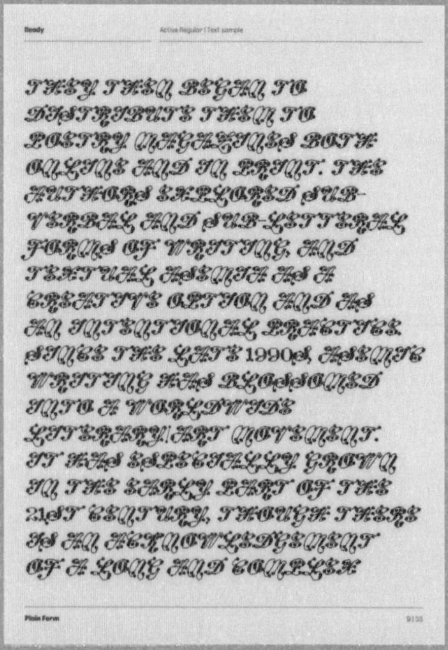

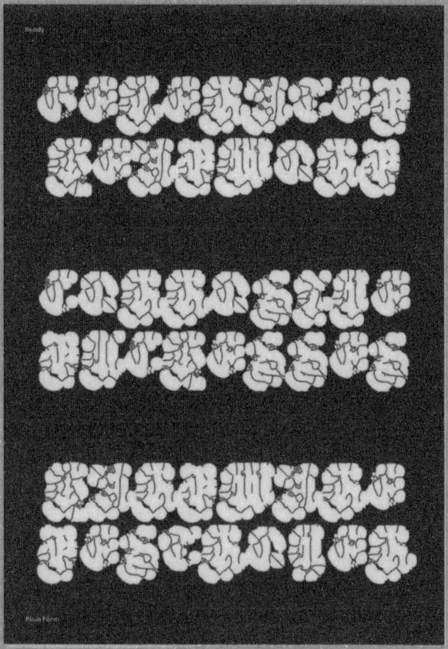

Plain Form
Typeface: Ready
Design: Lucas Descroix, Benjamin Dumond
—
plain-form.com @plain_form

240 pt

Support: Independent Type II

ADELLE SANS GEORGIAN DISPLAY SIZE

THIN 55PT

Rhythm & *Reaction*

THIN 50PT

Rhythm & *Reaction*

THIN 45PT

Rhythm & *Reaction*

THIN 35PT

Rhythm & *Reaction*

THIN 30/36PT

'Rhythm & Reaction' gets under the skin of a British love affair with American jazz. *Jazz first came to Britain as a visual and cultural style* — rather than as a musical form, writes John L. Walters.

ADELLE SANS GEORGIAN DISPLAY SIZE

HEAVY 55PT

Rhythm & *Reaction*

HEAVY 50PT

Rhythm & *Reaction*

HEAVY 45PT

Rhythm & *Reaction*

HEAVY 35PT

Rhythm & *Reaction*

HEAVY 30/36PT

'Rhythm & Reaction' gets under the skin of a British love affair with American jazz. *Jazz first came to Britain as a visual and cultural style — rather than as a musical form,* writes John L. Walters.

TYPETOGETHER WWW.TYPE-TOGETHER.COM © 2021 ALL RIGHTS RESERVED

ADELLE SANS GEORGIAN SPECIMEN

Magazine

UFO

Avistamiento

Las Juanas

kosmická loď

Starwars

Spaţiu

TYPETOGETHER WWW.TYPE-TOGETHER.COM © 2021 ALL RIGHTS RESERVED

TypeTogether
Typeface: Adelle Sans Georgian
Design: Veronika Burian, José Scaglione—Latin,
Jan Charvát, Akaki Radmadze, José Scaglione—Georgian
—
type-together.com @type_together

ADELLE SANS GEORGIAN SPECIMEN

საერთაშორისო

1786 წლის 8 ივნისს ნიუ-ორკში გაყიდვაში პირველად გამოიტანეს ნაყინი

🔊/👍 ინტერნეტი

ლამაზი

1 ₾ lari ლარი = 100 tetri თეთრი

ქართველები

MAGAZINE

შოთა რუსთაველის გამზირი ქუთაისი

ტიპოგრაფია

TYPETOGETHER WWW.TYPE-TOGETHER.COM © 2021 ALL RIGHTS RESERVED

ΚΥΚΛΟΣ
ΤΩΝ
ΛΑΒΔΑΚΙΔΩΝ

SOPHOCLES'
THEBAN
PLAYS

ΘΗΒΑΪΚΟΣ
ΔΡΑΜΑΤΙΚΟΣ
ΚΥΚΛΟΣ

OEDIPUS
REX
(TYRANNUS)

ΛΑΪΟΣ
ΙΟΚΑΣΤΗ
ΟΙΔΙΠΥΣ
ΚΡΕΟΝΤΑΣ
ΤΕΙΡΕΣΙΑΣ

LAIUS
JOCASTA
OEDIPUS
CREON
TIRESIAS

242 pt

ABCDEЄFGHIJKLM
NOPQRSTUVWXYZ
ÆƁƏŒMMNNWWC◎
ΑΒΓΔΕЄΖΗΘΙΚΛΜΝΞΞ
ΟΠΡΣΤΥΦΧΨΩϜϚ
.,;:!?{([#⊗*↔↕])}
“«†‡◦§¶$€№»”
&& 0123456789 ↑→←↓
MADE WITH LOVE BY
ATYPICAL.GR, 2021–2022

ATYPICAL
Typeface: Oedipus
Design: George Triantafyllakos, G Design Studio
—
atypical.gr

ΜΕΓΑΛΑ ΔΙΟΝΥΣΙΑ 401 Π.Χ.	FESTIVAL OF DIONYSUS 401 BC
ΑΝΤΙΓΟΝΗ	ANTIGONE
ΙΣΜΗΝΗ	ISMENE
ΘΗΣΕΑΣ	THESEUS
ΠΟΛΥΝΕΙΚΗΣ	POLYNICES
ΕΤΕΟΚΛΗΣ	ETEOCLES
ΘΗΒΑ	THEBES
ΑΘΗΝΑ	ATHENS
ΚΟΛΩΝΟΣ	COLONUS

Bureau Brut | Droulers Regular 02|16

80|84pt

However, I admitted the existence of the “MONSTER.”

Bureau Brut
Typeface: Droulers
Design: Bureau Brut
—
bureaubrut.com @bureau_brut

Bureau Brut | Droulers Regular 03|16

36|42pt

An immense track,
of dazzling whiteness,
marked the passage
of the animal

THE VOYAGE WAS
BEING ACCOMPLISHED
UNDER THE MOST
FAVOURABLE AUSPICES

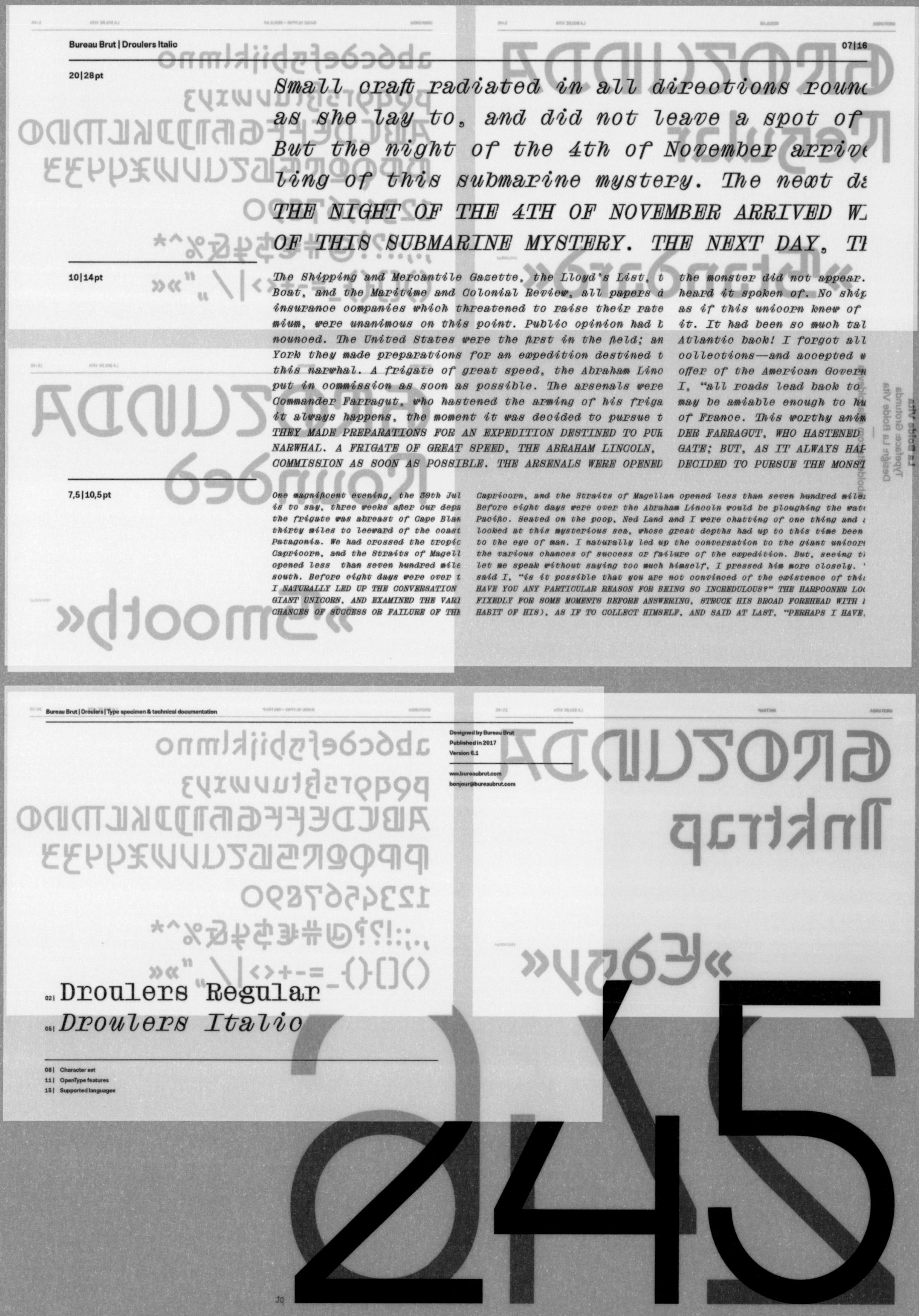
Bureau Brut | Droulers Italic

07|16

20|28pt

Small craft radiated in all directions round
as she lay to, and did not leave a spot of
But the night of the 4th of November arrive
ling of this submarine mystery. The next da
THE NIGHT OF THE 4TH OF NOVEMBER ARRIVED W
OF THIS SUBMARINE MYSTERY. THE NEXT DAY, TH

10|14pt

The Shipping and Mercantile Gazette, the Lloyd's List, t
Boat, and the Maritime and Colonial Review, all papers d
insurance companies which threatened to raise their rate
mium, were unanimous on this point. Public opinion had b
nounced. The United States were the first in the field; an
York they made preparations for an expedition destined t
this narwhal. A frigate of great speed, the Abraham Linc
put in commission as soon as possible. The arsenals were
Commander Farragut, who hastened the arming of his friga
it always happens, the moment it was decided to pursue t
THEY MADE PREPARATIONS FOR AN EXPEDITION DESTINED TO PUR
NARWHAL. A FRIGATE OF GREAT SPEED, THE ABRAHAM LINCOLN,
COMMISSION AS SOON AS POSSIBLE. THE ARSENALS WERE OPENED

the monster did not appear.
heard it spoken of. No ship
as if this unicorn knew of
it. It had been so much tal
Atlantic back! I forgot all
collections—and accepted w
offer of the American Govern
I, "all roads lead back to
may be amiable enough to hu
of France. This worthy anim
DER FARRAGUT, WHO HASTENED
GATE; BUT, AS IT ALWAYS HAP
DECIDED TO PURSUE THE MONST

7,5|10,5pt

One magnificent evening, the 30th Jul
is to say, three weeks after our depa
the frigate was abreast of Cape Blan
thirty miles to leeward of the coast
Patagonia. We had crossed the tropic
Capricorn, and the Straits of Magell
opened less than seven hundred mile
south. Before eight days were over t
I NATURALLY LED UP THE CONVERSATION
GIANT UNICORN, AND EXAMINED THE VARI
CHANCES OF SUCCESS OR FAILURE OF THE

Capricorn, and the Straits of Magellan opened less than seven hundred miles
Before eight days were over the Abraham Lincoln would be ploughing the wate
Pacific. Seated on the poop, Ned Land and I were chatting of one thing and a
looked at this mysterious sea, whose great depths had up to this time been
to the eye of man. I naturally led up the conversation to the giant unicorn
the various chances of success or failure of the expedition. But, seeing th
let me speak without saying too much himself, I pressed him more closely.
said I, "is it possible that you are not convinced of the existence of this
HAVE YOU ANY PARTICULAR REASON FOR BEING SO INCREDULOUS?" THE HARPOONER LOO
FIXEDLY FOR SOME MOMENTS BEFORE ANSWERING, STRUCK HIS BROAD FOREHEAD WITH H
HABIT OF HIS), AS IF TO COLLECT HIMSELF, AND SAID AT LAST, "PERHAPS I HAVE,

Bureau Brut | Droulers | Type specimen & technical documentation

Designed by Bureau Brut
Published in 2017
Version 6.1

www.bureaubrut.com
bonjour@bureaubrut.com

02| Droulers Regular
05| *Droulers Italic*

La Bolde Vita
Typeface: Grotunda
Design: La Bolde Vita
—
laboldevita.com @laboldevita

Support Independent Type II

GROTUNDA SAMPLE WATERFALL • REGULAR LA BOLDE VITA 13/45

‡ La lletra gòtica rotunda, gòtica rodona o gòtica de llibres és un tipus de lletra que pertany al grup d'estils cal·ligràfics sorgits en el període gòtic durant els segles segle xii — segle xv 40/48 pt

‡ La lletra gòtica rotunda, gòtica rodona o gòtica de llibres és un tipus de lletra que pertany al grup d'estils cal·ligràfics sorgits en el període gòtic durant els segles segle xii — segle xv (lletra gòtica), en oposició a l'escriptura «ANTIGA» 32/38 pt

TEXT SAMPLE https://ca.wikipedia.org/wiki/Rotunda

GROTUNDA BASIC GLYPHS • SMOOTH LA BOLDE VITA 16/45

abcdefghijklmno
pqqrsßtuvwxyz
ABCDEFFGHIJJKLMNO
PPQRSßTUVWXYYZZ
1234567890
,.;:!?¿@#€$¥&%^*
()[]{}_=-+‹›|/„"»«

GROTUNDA SAMPLE UPPERCASE • SMOOTH LA BOLDE VITA 19/45

REBEL = 1948 JULY
{UNIVERSAL LEAGUE}
CAFÉ GOLDSTEIN®
KOREA × JITTER ▸ K&J
C. 1700 — C. 500 FTR
BELIEF • SUBSTANCES
BOLSHEVISM.NEO ↗ 60/70 pt

GROTUNDA SAMPLE UPPERCASE • INKTRAP LA BOLDE VITA 29/45

REBEL = 1948 JULY
{UNIVERSAL LEAGUE}
CAFÉ GOLDSTEIN®
KOREA × JITTER ▸ K&J
C. 1700 — C. 500 FTR
BELIEF • SUBSTANCES
BOLSHEVISM.NEO ↗ 60/70 pt

Roumald

Regular, *Italic*,
Bold & *Bold-Italic*

nn *nn*

Roumald 245–145 pt 3

Groß
Inserate
Officer
Motel

Roumald 22 pt 6

LINN BOYD & MORRIS FULLER BENTON: father and son, each played crucial parts in the development of modern typefounding in the United

STATES AND THE WORLD. BOTH worked at the American Type Founders COMPANY which as a general policy did not promote or advertise the

IMPORTANCE OF INDIVIDUAL employees. Although MORRIS "sublimated his talents to the needs of a commercial type foundry," never-

THELESS "HIS SCORES OF REMARKABLY successful designs form the backbone of AMERICAN type design." At the same time, the mech-

248 pt

Erkin Karamemet
Typeface: Roumald
Design: Erkin Karamemet
—
erkinkaramemet.com @erkin_karamemet

Roumald
87 pt
4
Bold-Italic
Spezialisten
Italic
Klangfarben
Bold
Meisterhaus
Roman
Enthousiaste
Bold-Italic
Typografisch
Bold
Neutralisiert
© Erkin Karamemet Studio

The Shining 1980
① by STEPHEN KING?
psychological horror
starring Room 237
❷ * Jack Nicholson!
»Stanley Kubrick« ↙
Distributed & Directed
№7→Warner Bros.®
BUDGET: 19 million
↳ Released UK 1980
Running 146 min ¶

Standard

Slim
Neutral
Median
Bulky

Mono

Slim
Neutral
Median
Bulky

Wide

Slim
Neutral
Median
Bulky

Italic

Slim
Neutral
Median
Bulky

Italic

Slim
Neutral
Median
Bulky

Italic

Slim
Neutral
Median
Bulky

250

pt

Gradient
Typeface: PolySans
Design: Milos Mitrovic
—
wearegradient.net @gradient.type

Support Independent Type II

*Big
Wtf ~
Љубав!
% Posh
Buck $

18

3⅞ ®

25

Gradient
Typeface: Okay! Serif
Design: Milos Mitrovic
—
wearegradient.net @gradient.type

Grilli Type

Maru Emojis

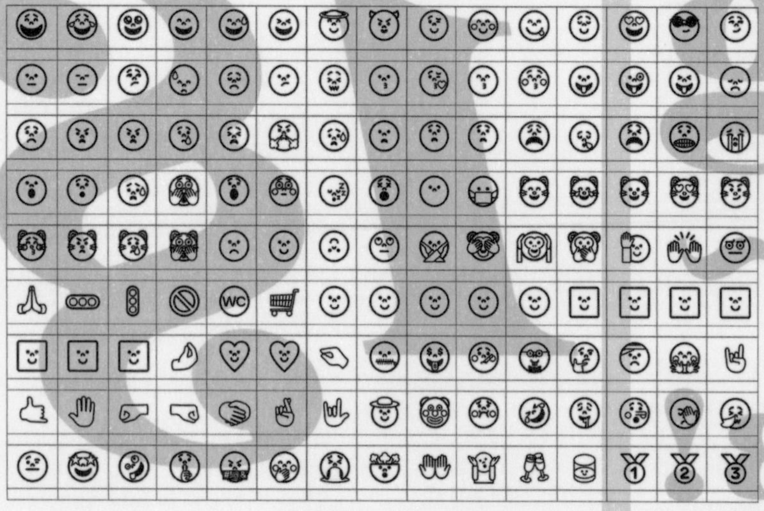

GT Maru by Thierry Blancpain / © 2021 Grilli Type / www.grillitype.com 13/42

Grilli Type

GT Maru Light 125pt, Tracking -30pt

Mount Fuji

GT Maru Light 170pt, Tracking -60pt

Shibuya

GT Maru Light 190pt, Tracking -50pt

Minato

GT Maru by Thierry Blancpain / © 2021 Grilli Type / www.grillitype.com 16/42

Grilli Type

Bento
BOX

Grilli Type

Grilli Type

Famed for its beautiful trees that turn bright yellow in autumn, cafes, restaurants, and a large replica of Michelange-lo's David outside of the Papas building.

Grilli Type

The southern half of Shin-juku Station, including the New South En-

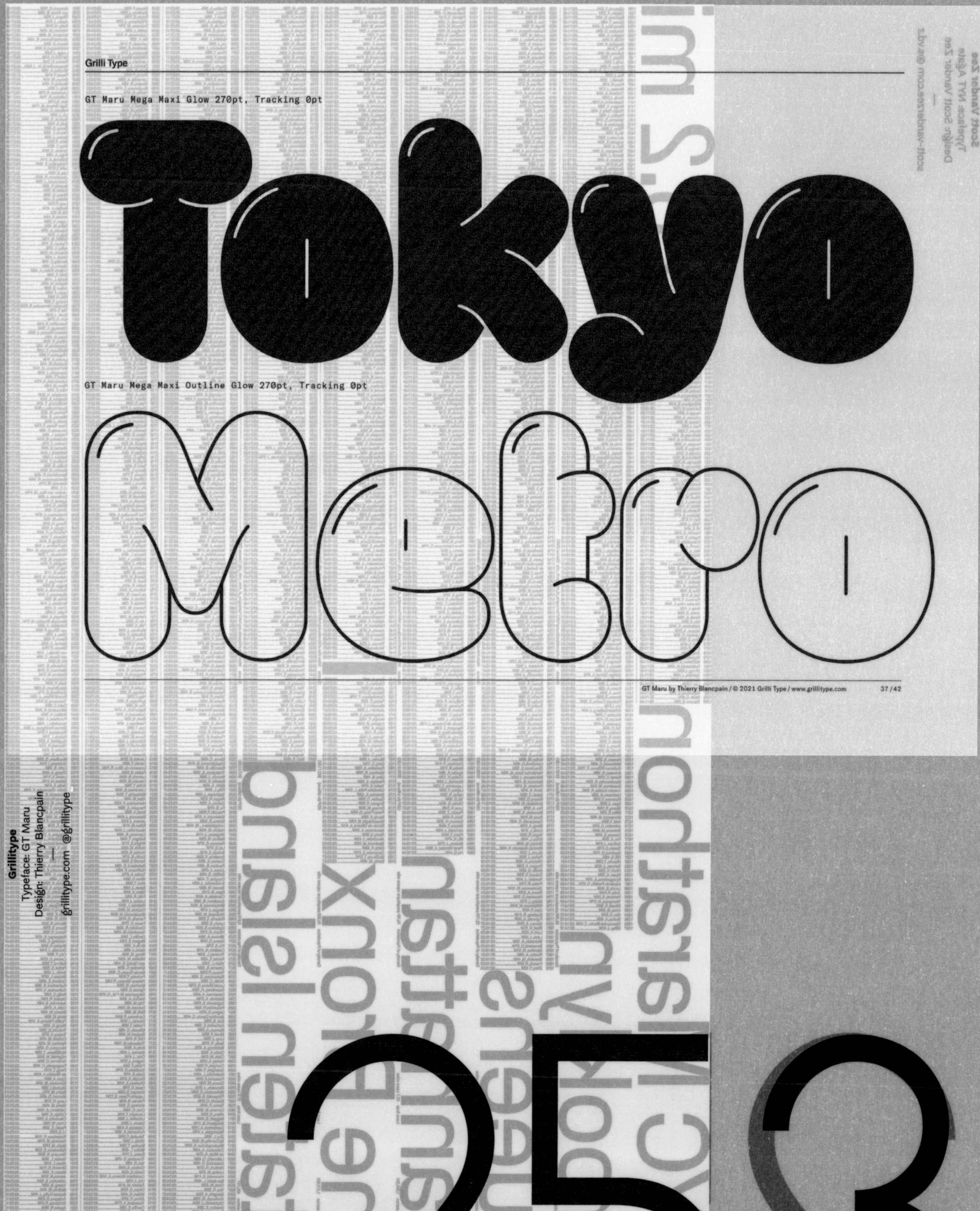

Grillitype
Typeface: GT Maru
Design: Thierry Blancpain
—
grillitype.com @grillitype

23 pt

Scott Vander Zee
Typeface: NYT Agate
Design: Scott Vander Zee
—
scott-vanderzee.com @s.vd.z

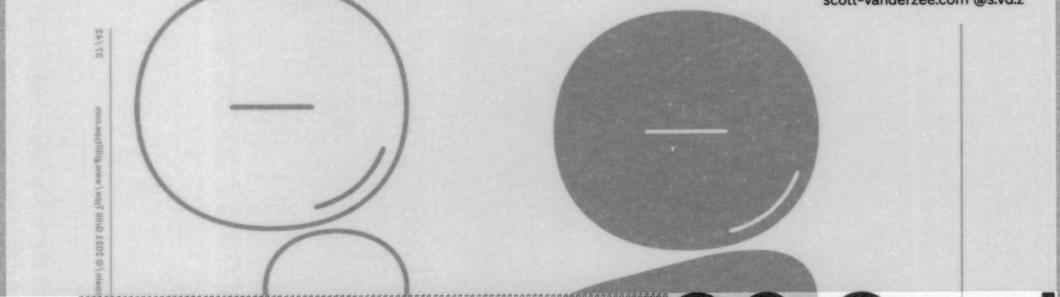

(New York Times)
ABCDEFGH
IJKLMNOPQRST
UVWXYZ WMM
0123456789.,:;●/
abcdefgh
ijklmnopqrst
uvwxyz
The 2024 TCS
New York City
Marathon
Sun., 3 Nov. 2024
[Agate Regular]

pt 255

Optimo
Typeface: Void
Design: Malte Bentzen
—
optimo.ch @optimofoundry

Striated
NAME
CREATOR
Anna Dudina
RELEASE
2022
Specimen & Technical Documentation
SHRIFTOVIK
Spread
SHRIFTOVIK
FOUNDRY
TYPE DESIGN
EXPERIMENTS
Road to Venice Type
Typeface: Striated
Design: Road to Venice Type
—
r-vtype.com @r.v.type
Bold / Bold Italic
(So full of 60)
shapes is fancy
ENOUGH
#Nuf 12
Main7
→LIVE, BRUTESS, LIVE!
Bring them with triumph
to their house; 1234
»BRUTESS!«
Silence, silence!
She speaks 12
NAME HER NOT
“Indeed”, they
is [not] fit 25
Technical Documentation
5
257

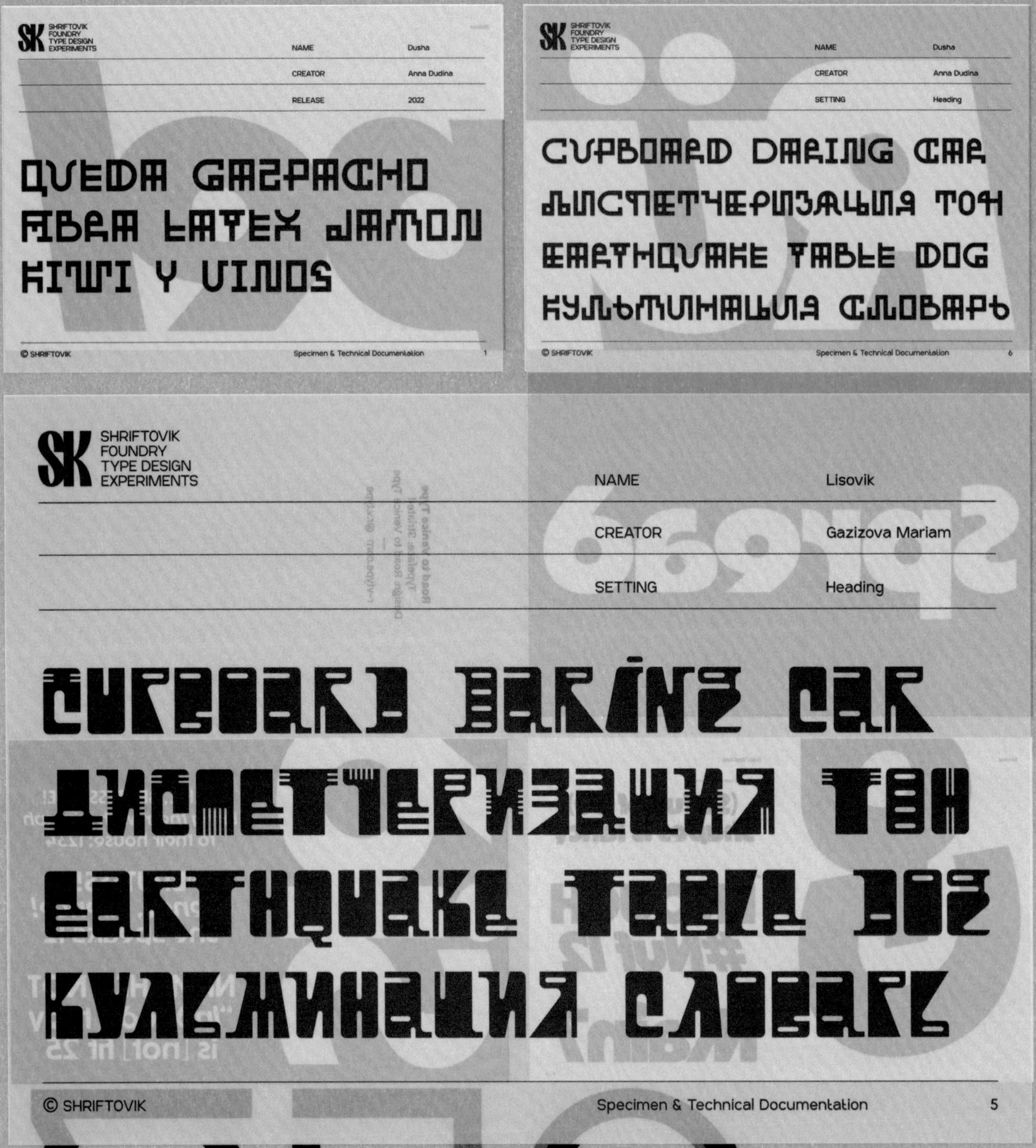
SK SHRIFTOVIK FOUNDRY TYPE DESIGN EXPERIMENTS
NAME Dusha
CREATOR Anna Dudina
RELEASE 2022
© SHRIFTOVIK
Specimen & Technical Documentation 1
SK SHRIFTOVIK FOUNDRY TYPE DESIGN EXPERIMENTS
NAME Dusha
CREATOR Anna Dudina
SETTING Heading
© SHRIFTOVIK
Specimen & Technical Documentation 6
SK SHRIFTOVIK FOUNDRY TYPE DESIGN EXPERIMENTS
NAME Lisovik
CREATOR Gazizova Mariam
SETTING Heading
CUPBOARD DARING CAR
ДИСПЕТЧЕРИЗАЦИЯ ТОН
EARTHQUAKE TABLE DOG
КУЛЬМИНАЦИЯ СЛОВАРЬ
© SHRIFTOVIK
Specimen & Technical Documentation 5

258 pt

SHRIFTOVIK FOUNDRY TYPE DESIGN EXPERIMENTS
NAME
Dusha
CREATOR
Anna Dudina
CHARACTER SET
Basic Lowercase
© SHRIFTOVIK
Specimen & Technical Documentation
4
SHRIFTOVIK FOUNDRY TYPE DESIGN EXPERIMENTS
NAME
Lisovik
CREATOR
Gazizova Mariam
CHARACTER SET
Basic Uppercase
© SHRIFTOVIK
Specimen & Technical Documentation
3
Shriftovik
Typeface: Lisovik, Dusha
Design: Gazizova Mariam, Anna Dudina
shriftovik.com @shriftovik

Skrr Type
Typeface: Alna
Design: Alff Rosine
—
skrr-type.com @skrr_type

Source Type
Typeface: Chroma
Design: Selina Bernet
—
sourcetype.com @sourcetype

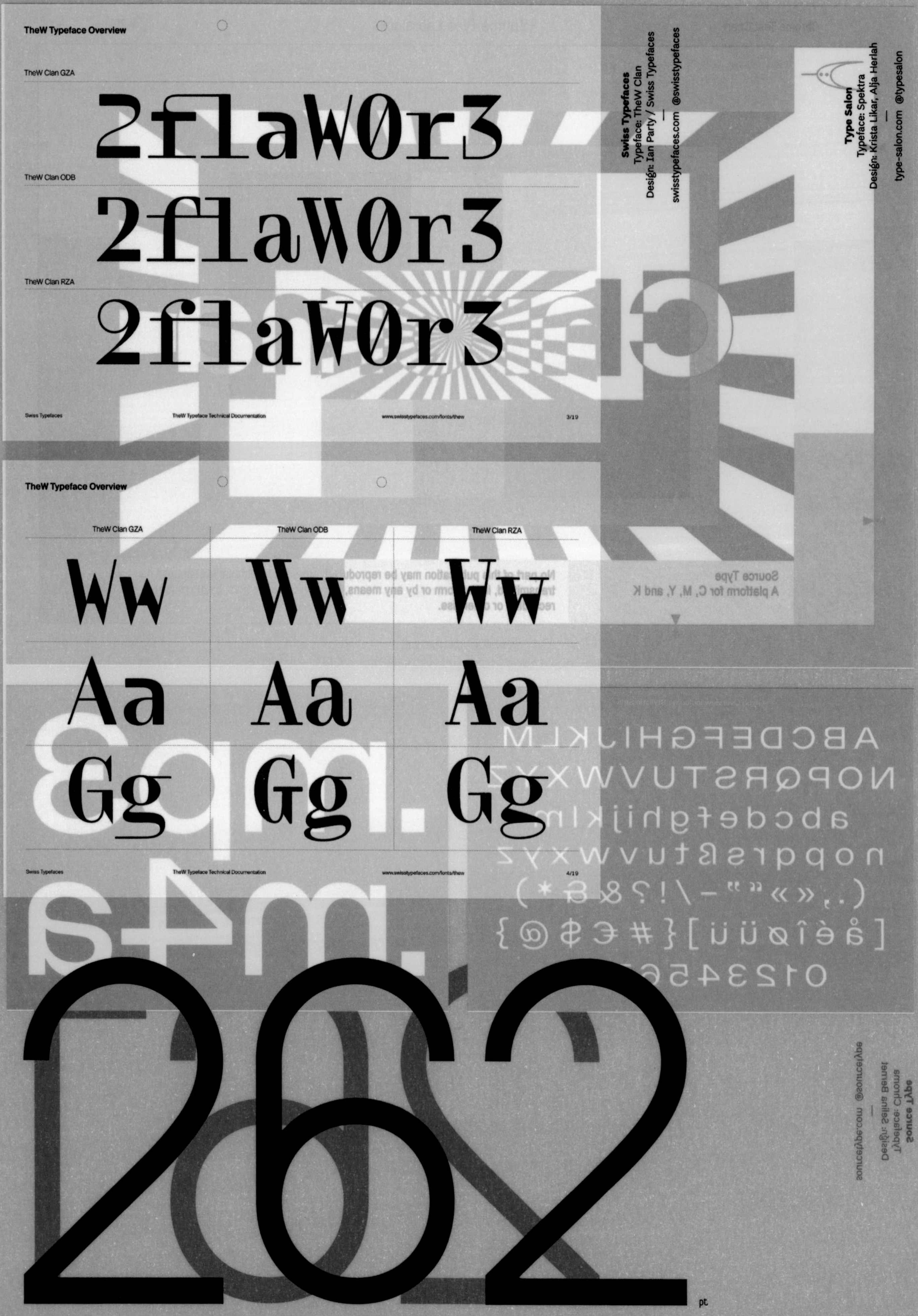
TheW Typeface Overview
TheW Clan GZA
2flaW0r3
TheW Clan ODB
2flaW0r3
TheW Clan RZA
2flaW0r3
Swiss Typefaces
TheW Typeface Technical Documentation
www.swisstypefaces.com/fonts/thew
3/19
TheW Typeface Overview
TheW Clan GZA
TheW Clan ODB
TheW Clan RZA
Ww
Ww
Ww
Aa
Aa
Aa
Gg
Gg
Gg
Swiss Typefaces
TheW Typeface Technical Documentation
www.swisstypefaces.com/fonts/thew
4/19
Swiss Typefaces
Typeface: TheW Clan
Design: Ian Party / Swiss Typefaces
swisstypefaces.com @swisstypefaces
Type Salon
Typeface: Spektra
Design: Krista Likar, Alja Herlah
type-salon.com @typesalon
Source Type
A platform for C, M, Y, and K
262
pt

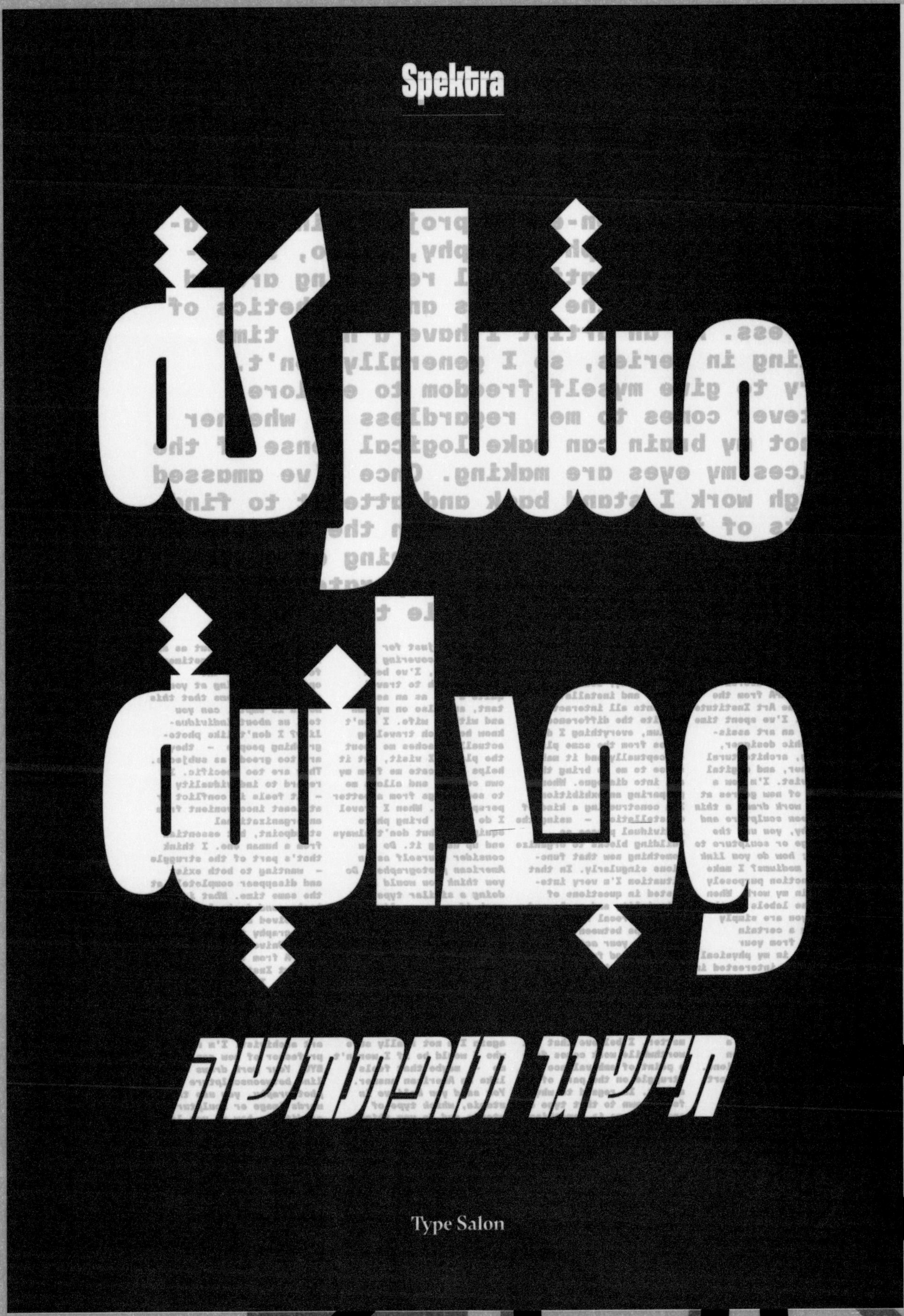
Spektra
مشاركة
وجدانية
Type Salon

76pt

beautiful

15pt
18pt

It's a fairly open-ended project, incorporating elements of photography, video, sculpture, and installation all revolving around questions about the ideals and aesthetics of progress. As an artist I have a hard time working in series, so I generally don't. I try to give myself freedom to explore whatever comes to me, regardless of whether or not my brain can make logical sense of the choices my eyes are making. Once I've amassed enough work I stand back and attempt to find points of intersection between the individual pieces. I see my work functioning as a loose constellation of somewhat disparate parts building a complicated single tone. These

6pt
8pt

What is your background (school, works)? I received a BFA in photography from Brigham Young University, and then an MFA from the School of The Art Institute of Chicago. I've spent time working as an art assistant, graphic designer, missionary, architectural photographer, and digital art archivist. I'm now a professor of new genres at BYU. *Your work draws a thin line between sculpture and photography, you use the words image or sculpture to credit it; how do you link these two mediums?* I make the distinction purposely slippery in my work. When using those labels on the Internet you are simply requesting a certain perception from your viewer, but in my physical work I'm also interested in a similar kind of slippage. I'm interested in how a photograph changes when treated as a sculptural object and likewise, how the understanding of a sculpture changes when reduced to two dimensions. Overall, in making my art I'm not concerned with finality. One of the advantages of working digitally is the ability to endlessly revisit and revise work without ever having to ultimately commit. So even after something has been publicly exhibited in a particular form, I'm still open to letting it evolve and exist simultaneously in multiple states. *How are they working together in an exhibition space?* In my shows I generally have photography, sculpture, video, and installation elements all interacting. Despite the differences in medium, everything I do comes from the same place conceptually and it makes sense to me to bring them all into dialogue. When preparing an exhibition, I'm constructing a kind of constellation – using the individual pieces as building blocks to organize something new that functions singularly. In that situation I'm very interested in questions of materiality and scale and the reciprocal relationships between pieces. *What are your actual thoughts and feelings concerning the non places, or anonymous spaces you photographed?* I'm still figuring that out. I think once I do, I'll be done with it as a subject matter. I believe that all worthwhile work comes from a point of ambivalence or struggle on the part of the artist. In regard to why I feel drawn to that type of space, there is something very appealing to me about blandness, order, and the aesthetics of efficiency. Part of me wants to believe wholeheartedly in the promises and ideals of modernism, but I was born too late for that kind of thinking. *It seems that you have travelled a lot, was it mostly for photographic purpose or just for the wish of discovering new lands?* Yeah, I've been lucky enough to travel quite a bit as an assistant, and also on my own and with my wife. I don't know how much traveling actually teaches me about the places I visit, but it helps dislocate me from my own context and allows me to see things from a better perspective. When I travel I do always bring photo equipment, but don't always end up using it. *Do you consider yourself as an American photographer? Do you think you would be doing a similar type of work if you were living in Paris?* No, I don't think of myself as an American photographer. In my work I'm trying to escape specificity, including the connotations of any particular location or culture. I would hope I'd be making the same work if I were from Paris, but then again I'm not really sure who I would be if I weren't me – maybe that feels like an American answer. *You said you believe in utopia, which type of utopia and do you think it's still possible? Is Art a utopia?* A part of me is drawn to the kind of all-or-nothing thinking that leads to manifestoes and compound-based communities, but I also know how those things always turn out. I think for the time being utopia is a personal thing, and maybe art can facilitate that, but as an institution art sometimes feels like just the opposite. *Looking at your pictures it seems that this world is empty, can you tell us about individuality?* I don't like photographing people – they are too greedy as subjects. They are too specific. In regard to individuality – it feels in conflict or at least inconvenient from an organizational standpoint, but essential from a human one. I think that's part of the struggle – wanting to both exist and disappear completely at the same time. *What is your background (school, works)?* I received a BFA in photography from Brigham Young University, and then an MFA from the School of The Art Institute of Chicago. I've spent time working as an art assistant, graphic designer, missionary, architectural photographer, and digital art archivist. I'm now a professor of new genres at BYU. *Your work draws a thin line betweensculpture and photography, you use the words image or sculpture to credit it; how do you link these two mediums?* I make the distinction purposely slippery in my work. When using those labels on the Internet you are simply requesting a certain perception from your viewer, but in my physical work I'm also interested in a similar kind of slippage. I'm interested in how a

pt

Support Independent Type II

TYPE.WELTKERN® TWK Everett Mono Extrabold + Italic
CH
ZÜRICH
salt (sôlt) n.
1 A colorless or white crystalline solid used extensively in ground or granulated form as a food seasoning and preservative
2 An element that gives flavor or zest.
3 Sharp, lively wit.
4 A mineral sharing definitive characteristics with Utah's capital city.
TYPE.WELTKERN® TWK Everett Mono Super + Italic
MAN
INTERESTED REVISITING SCULPTURES
I make the distinction purposely slippery in my work. (…) but in my physical work I'm also interested in a similar kind of slippage. I'm interested in how a photograph changes when treated as a sculptural object and likewise, how the understanding of a sculpture changes
WELTKERN® Typefaces
Typeface: TWK Everett Mono
Design: Nolan Paparelli
—
weltkern.com @weltkern
TYPE.WELTKERN® TWK Everett Mono Regular + Italic 12/30
But the beauty of his work lies in the way it isn't alienating in the slightest. We first fell for his compositions when
"[…] we were always traveling to some superlative location — the biggest, the widest, the greatest — and I got really interested in the in-between places that we passed through: the nondescript, transitory spaces like subway systems, airports, parking garages, and hotels. Spaces that are meant to be legible regardless of the language, and where the aesthetics are governed by function." 21.01.15
TYPE.WELTKERN® TWK Everett Mono Light + Italic 9/30
Digitals
Platform
Internet
Systemic
Personal
Manifest
Visually
Feelings
He has a very endearing way of destroying your perception of photography & making you question the ultimate function and actual capabilities of the medium. Other than
Calm
Elysian
Quiet Place
265

Type of Feeling
Typeface: Elysian
Design: Type of Feeling
typeoffeeling.com @typeoffeeling
typeoffeeling.com
Elysian Light
Elysian Regular
Elysian Medium
Elysian Bold
CH
ZÜRICH
salt (sôlt) n.
1 A colorless or white crystalline solid used extensively in ground or granulated form as a food seasoning and preservative
2 An element that gives flavor or zest.
3 Sharp, lively wit.
4 A mineral sharing definitive characteristics with Utah's capital city.
234 PT Regular
Calm
But the beauty of his work lies in the way it isn't alienating in the slightest. We first fell for his compositions when
ABCČĆDĐEFGH
IJKLMNOPQRSŠ
TUVWXYZŽab
cčćdđefghijklm
nopqrsštuvwxyzž
1234567890
ch ct ff ffh fj gi st
Quiet Place
Being creatively & divinely inspired – utterly in a state of blissful delight & perfection.
Elysian is a Transitional Serif typeface inspired by the feeling of being creatively & divinely inspired – utterly in a state of blissful delight & perfection. As an ode to the paradisiacal Elysium from Greek mythology, this font has embellished terminals & features, sharp & compelling Serifs, and classically-styled numerals.
266
pt
Support Independent Type II

typeoffeeling.com 2

45 PT

Satori Thin

Thin — Satori Thin

Satori Light

Light — Satori Light

Satori Regular

Regular — Satori Regular

Satori Medium

Medium — Satori Medium

Satori Bold

Bold — Satori Bold

Type of Feeling
Typeface: Satori
Design: Type Of Feeling
—
typeoffeeling.com @typeoffeeling

typeoffeeling.com 5

50 PT / 52 LH

ABCČĆDĐEFGH
IJKLMNOPQRSŠ
TUVWXYZŽab
cčćdđefghijklm
nopqrsštuvwxyzž

'?"!"(%)[#]{@}/&\<
+÷×=>®©$€£¥¢:;,.*

1234567890

Glyphs — Satori Thin

JJJ KK PP RRR ULI

Alt Glyphs — Satori Thin

285 PT Thin

Charm

typeoffeeling.com 9

ABCČĆDĐEFGH
IJKLMNOPQRSŠ
TUVWXYZŽab
cčćdđefghijklm
nopqrsštuvwxyzž

'?"!"(%)[#]{@}/&\<
+÷×=>®©$€£¥¢:;,.*

1234567890

Glyphs — Satori Bold

JJJ KK PP RRR ULI

Alt Glyphs — Satori Bold

typeoffeeling.com 4

67 PT / 67 LH / Bold

Enlightened Growth

22 PT / 25 LH / Medium

Inspiration from the circular journey of life and self-discovery, contrasted with the flat planes to represent the moment of realization and awakening.

17 PT / 20 LH / Regular

Satori is a geometric Sans Serif typeface inspired by feelings of sudden enlightenment or spiritual awakening. The type design blends inspiration from the circular journey of life and self-discovery, contrasted with the flat planes to represent the moment of realization and awakening.

11 PT / 15 LH / Light & Thin

Satori is a geometric Sans Serif typeface inspired by feelings of sudden enlightenment or spiritual awakening. The type design blends inspiration from the circular journey of life and self-discovery, contrasted with the flat planes to represent the moment of realization and awakening.

Satori is a geometric Sans Serif typeface inspired by feelings of sudden enlightenment or spiritual awakening. The type design blends inspiration from the circular journey of life and self-discovery, contrasted with the flat planes to represent the moment of realization and awakening.

Support Independent Type II

Pivot Grotesk
Black, Black Italic
120/100 –8
Earth
Material
32/34.4 0
To suppose that the eye with all its inimitable contrivances for adjusting the focus to different distances, for admitting
12/15 +15
To suppose that the eye with all its inimitable contrivances for adjusting the focus to different distances, for admitting different amounts of light, and for the correction of spherical and chromatic aberration, could have been formed by natural selection, seems, I freely confess, absurd in the highest degree. When it was first said that the sun stood still and the world turned round, the common sense of mankind declared the doctrine false; but the old saying of Vox populi, vox Dei, as every philosopher knows, cannot be trusted in science. Reason tells me, that if numerous gradations from a simple and imperfect eye to one complex and perfect can be shown to exist, each
8/9.6 +30
To suppose that the eye with all its inimitable contrivances for adjusting the focus to different distances, for admitting different amounts of light, and for the correction of spherical and chromatic aberration, could have been formed by natural selection, seems, I freely confess, absurd in the highest degree. When it was first said that the sun stood still and the world turned round, the common sense of mankind declared the doctrine false; but the old saying of Vox populi, vox Dei, as every philosopher knows, cannot be trusted in science. Reason tells me, that if numerous gradations from a simple and imperfect eye to one complex and perfect can be shown to exist, each grade being useful to its possessor, as is certainly the case; if further, the eye ever varies and the variations be inherited, as is likewise certainly the case; and if such variations should be useful to any animal under changing conditions of life, then the difficulty of believing that a perfect and complex eye could be formed by natural selection, though insuperable by our imagination, should not be considered as subversive of the theory. How a nerve comes to be sensitive to light, hardly concerns us more than
© 2024 Nouvelle Noire — www.nouvellenoire.ch
5 of 13
Nouvelle Noire
Typeface: Pivot Grotesk
Design: Anton Studer
—
nouvellenoire.ch @nouvellenoire_ch
Pivot Grotesk
Regular, Italic
Universal
Motions?
To suppose that the eye with all its inimitable contrivances for adjusting the focus to different distances, for admitting diffe-
To suppose that the eye with all its inimitable contrivances for adjusting the focus to different distances, for admitting different amounts of light, and for the correction of spherical and chromatic aberration, could have been formed by natural selection, seems, I freely confess, absurd in the highest degree. When it was first said that the sun stood still and the world turned round, the common sense of mankind declared the doctrine false; but the old saying of Vox populi, vox Dei, as every philosopher knows, cannot be trusted in science. Reason tells me, that if numerous gradations from a simple and imperfect eye to one complex and perfect can be shown to exist, each grade being useful to its possessor, as is certainly the case; if
© 2024 Nouvelle Noire — www.nouvellenoire.ch
Pivot Grotesk
Thin, Thin Italic
Returning
Histories
To suppose that the eye with all its inimitable contrivances for adjusting the focus to different distances, for admitting different
To suppose that the eye with all its inimitable contrivances for adjusting the focus to different distances, for admitting different amounts of light, and for the correction of spherical and chromatic aberration, could have been formed by natural selection, seems, I freely confess, absurd in the highest degree. When it was first said that the sun stood still and the world turned round, the common sense of mankind declared the doctrine false; but the old saying of Vox populi, vox Dei, as every philosopher knows, cannot be trusted in science. Reason tells me, that if numerous gradations from a simple and imperfect eye to one complex and perfect can be shown to exist, each grade being useful to its possessor, as is certainly the case; if further, the eye ever
© 2024 Nouvelle Noire — www.nouvellenoire.ch

Approximate Type © 2023 — Skanner Specimen, Ligatures, Bold

105 pt

HELIOTEC!
Sisdematic

74 pt

HOSTESS 157?
Photosynthesis

50 pt

HOOVER MANEUVER
Oprassive Infactions

30 pt

REAL HOOPERS SPORTSCENTER 360
Hazmatsuits & Antiviral Technologies

20 pt

INTRAVENOUS UPHOLSTERY, 404 OPERATOR ERROR?
Hangtime Atlas, Thesaurus, StarTrek, Minority Report!

Approximate Type © 2023 — Skanner Specimen

Skanner Skanner
Skanner Skanner
Skanner Skanner
Skanner Skanner
Skanner Skanner
Skanner Skanner

Commercial cargo vessels, glitchy VHS tapes, electric hoverbikes and sci-fi user interfaces all seem to share a distinct aesthetic that is hardcoded into Skanner's geometric, prosthetic exoskeleton. Like items on a conveyor belt, Skanner's letterforms follow each other in precise, rhythmic succession. Its squared off bowls allow for barcode-tight tracking while its horizontal terminals suggest a plethora of ligatures, of which a large variety is included in the typeface—123 of their kind to be exact!

While Skanner is not based on any specific reference, Eurostile springs to mind. In comparison, Skanner introduces a constructed and rigorous design language as well as deep ink traps that contribute a subtle sense of warmth to the typeface.

Released: 2023
Designer: Kasper Pyndt Rasmussen

Weights: 6 — Styles: 12 — Variable font
Availability: Desktop, Web, App

Approximate Type
Typeface: Skanner
Design: Kasper Pyndt Rasmussen
approxtype.com @approx.type

270 pt

Approximate Type
Typeface: Dalat
Design: Kasper Pyndt Rasmussen
—
approxtype.com @approx.type

Approximate Type © 2021 — Dalat Specimen, headlines (Bold)

EXPLICIT!
Traffic Jam

100 pt

15 CONFLICTS
Voyager Influx?

70 pt

DANH VO, ARTIST?
Immediate Aftermath

50 pt

218 CALLIGRAPHIC REFERENCES
Craftsmanship & Self-governance

30 pt

FRANCIS FORD COPPOLA OR OLIVER STONE?!
Experts in disagreement over Vietnam-flicks

20 pt

Approximate Type © 2021 — Dalat Specimen, running text (Light)

THERE ARE 29 LETTERS IN THE VIETNAMESE alphabet. There are four to six tones, which are marked in the IPA as suprasegmentals following the phonemic value. It uses all the letters of the ISO basic Latin alphabet except for F, J, W, and Z. The aforementioned letters are only used to write loanwords, languages of other ethnic (...)

22 pt

THERE ARE 29 LETTERS IN THE VIETNAMESE ALPHABET. THERE are four to six tones, which are marked in the IPA as suprasegmentals following the phonemic value. It uses all the letters of the ISO basic Latin alphabet except for F, J, W, and Z. The aforementioned letters are only used to write loanwords, languages of other ethnic groups in the country based on vietnamese phonetics to differentiate the meanings or even vietnamese dialects, for example: dz or z for southerner pronunciation of v in standard vietnamese. The Vietnamese (...)

18 pt

THERE ARE 29 LETTERS IN THE VIETNAMESE ALPHABET. THERE ARE FOUR TO six tones, which are marked in the IPA as suprasegmentals following the phonemic value. It uses all the letters of the ISO basic Latin alphabet except for F, J, W, and Z. The aforementioned letters are only used to write loanwords, languages of other ethnic groups in the country based on vietnamese phonetics to differentiate the meanings or even vietnamese dialects, for example: dz or z for southerner pronunciation of v in standard vietnamese. The Vietnamese alphabet, chữ Quốc ngữ, literally "National language script", is the modernwriting system for the Vietnamese language. It uses the Latin script based on Romance languages, in particular, the Portuguese alphabet, with some digraphs (...)

12 pt

THERE ARE 29 LETTERS IN THE VIETNAMESE alphabet. There are four to six tones, which are marked in the IPA as suprasegmentals following the phonemic value. It uses all the letters of the ISO basic Latin alphabet except for F, J, W, and Z. The aforementioned letters are only used to write loanwords, languages of other ethnic groups in (...)

THERE ARE 29 LETTERS IN THE VIETNAMESE ALPHABET. THERE are four to six tones, which are marked in the IPA as suprasegmentals following the phonemic value. It uses all the letters of the ISO basic Latin alphabet except for F, J, W, and Z. The aforementioned letters are only used to write loanwords, languages of other ethnic groups in the country based on vietnamese phonetics to differentiate the meanings or even vietnamese dialects, for example: dz or z for southerner pronunciation of v in standard vietnamese. The Vietnamese alphabet, chữ Quốc ngữ, literally "National language script", is the modern writing system for the Vietnamese language. It uses the Latin script (...)

10 / 7 pt

Approximate Type © 2021 — Dalat Specimen, description

Dalat Light
Dalat Regular
Dalat Bold

The starting point of the project was a trip to Vietnam and a subsequent curiosity about their prolific use of Cooper Black for street signage—ironically an archetypical American typeface. Foreign conflicts and colonisation are deeply entangled in the story of Vietnam which results in a very multi-facetted visual culture. The typeface Dalat is an unlikely hybrid and synthesis of these varying references; French Art Deco and Russian Constructivism comes across in the geometric construction of the letters, while the typeface's soft serifs are derived from Chinese/Vietnamese calligraphy (thu phap) whose terminals tend to thicken and pool. The assemblage of these styles results in a typeface that points East as well as West and finds its home in both contexts. With its decorative and playful characteristics, Dalat is mainly suitable for large text-sizes while maintaining good legibility in smaller sizes as well.

Released: 2020
Designer: Kasper Pyndt Rasmussen
Weights: 3 — Styles: 3
Availability: Desktop, Web, App

— VII —

Index

272 pt

273

D

E

F

G

H

N

O

P

Q

280

pt

Thank you to everyone who backed *Support Independent Type II*—from those who preordered the book to all who champion independent type.
Your passion and dedication inspire us and make this project possible.
Together, we celebrate creativity, individuality, and the vibrant culture of typography.

Fonts used:
Di Grotesk, Tor Grotesk, Tor Grotesk Mix
(Threedotstype)

Great books deserve great paper. Thanks to Reflex GmbH & Co. KG
for providing us with their finest selection—setting the perfect stage for design,
texture, and print excellence!
reflex-paper.com

This edition was printed on a carefully curated selection of fine papers:

Dust jacket: IBO One, 60 gsm
Cover & back cover: EFALIN 108 high white, new linen, 350 gsm
Inside pages: A mix of Zeta brilliant wove (120 gsm), IBO One (60 gsm), and Reflex
Transparent White (100 gsm) to enhance the reading and tactile experience.

Colophon
Slanted Publishers UG
(haftungsbeschränkt)
Nördliche Uferstraße 4–6
76189 Karlsruhe
Germany
T +49 (0) 721 85148268
info@slanted.de
slanted.de
@slanted_publishers

ISBN: 978-3-948440-82-4
1st edition 2025

Editors & Design: Lars Harmsen, Marian Misiak, Pola Małaczewska
Publishing Direction: Lars Harmsen, Julia Kahl
Production Management: Julia Kahl
Proofreading: Julia Kahl, Juliane Lipp
Printing: Stober Medien, Germany
Bookbinding: Spinner, Germany

The German National Library lists this publication in the German National Bibliography; detailed bibliographic data is available on the Internet at dnb.de.

About
Slanted Publishers is an independent design, publishing and media house founded in 2014 by Lars Harmsen and Julia Kahl. They publish the award-winning print magazine *Slanted* biannually featuring global design and culture. Since 2004, the daily blog *www.slanted.de* highlights international design and showcases inspiring video interviews. Slanted Publishers initiates and creates publications, focusing on contemporary design and visual culture, working closely with editors and authors to produce outstanding publications with meaningful content and high quality. Slanted was born from great passion and has made a name for itself across the globe. Its design is vibrant and inspiring—its philosophy open-minded, tolerant, and curious.

288

pt